THOMAS BOLTON
& SONS LIMITED
1783-1983

THOMAS BOLTON & SONS LIMITED

1783~1983

the bi-centenary history of a major copper and brass manufacturer

JOHN MORTON

MOORLAND PUBLISHING

ISBN 0 86190 056 1

Reproduced from copy supplied, printed and bound in Great Britain by Billing and Sons Limited, Worcester for Moorland Publishing Company Ltd, 9-11 Station Street, Ashbourne, Derbyshire DE6 1DE, England

Contents

Introduction

The copper works of Thomas Bolton and Sons Ltd, come upon the traveller unexpectedly, as he journeys along the winding A52 from Stoke-on-Trent to Ashbourne in Derbyshire. A sharp turn in the road at Froghall reveals the extensive factory buildings of the company filling the picturesque valley of the Churnet. The questions 'Who was Thomas Bolton and how did his copper works come to be here in this seemingly remote corner of Staffordshire?', must have arisen in the minds of many.

The curious traveller will be surprised to find that the answers to these questions take him back to the very earliest days of the Industrial Revolution in Britain. The firm now known as Thomas Bolton & Sons Ltd can trace its origins back two hundred years to when the Bolton family established a business in Birmingham. At that time, in the 1780s, Birmingham and the Black Country towns of Walsall, Wednesfield, Wolverhampton and Bilston were becoming leading centres for the manufacture of metal hardware of all kinds. The family fortunes expanded with those of the town. The Boltons progressed from a plater's workshop to a hardware merchant's warehouse with an international reputation by the early nineteenth century. Under the first Thomas Bolton, a steam-powered industrial establishment was set up in 1838, which began the same business of copper and brass fabrication which is still followed today at Froghall.

A decisive turning point came in 1852 when Thomas Bolton bought the wire and tube mill at Oakamoor near Froghall in Staffordshire. In the half century which followed Boltons became one of the leading houses in the copper and brass trade, at a time when British industry, and Birmingham industry in particular, dominated the world's markets for such products. They provided copper wire for the first transatlantic telegraph cable in 1857, and for the more successful second and third cables in 1865 and 1866. Under the leadership of Thomas Bolton's elder son, Alfred, the firm made further important contributions to the new age of electricity in the manufacture of high conductivity copper, the continuous drawing of fine wire and the making of copper wire in very long lengths for electrical transmission. Alfred himself played a part in the very first electric lighting companies, and in the early aluminium industry. The firm extended its activities to Widnes where Boltons began to smelt and refine their own copper for the first time in 1881. Froghall was built in 1890 to give more refinery capacity, and in 1894 the Sutton rolling mill at St Helens was acquired.

Thomas Bolton and Sons, as the firm became known after Thomas Bolton's death in 1853, was a medium-sized firm typical of those which were the backbone of British industry during its Victorian hey-day. Curiously, they have received little credit from historians for their contribution, although the reputation built up in the nineteenth century has been maintained in the trade till the present day. In part this lack of interest may be due to the benign neglect by historians which has been the fate of the non-ferrous metal industries when compared with the historical interest in iron and steel. It may also be due to the nature of the fabricator's business. His name never appears before the eyes of the public, unlike that of the manufacturer who uses his products in a finished article. Many people will be familiar with the names of cable manufacturers or telephone companies, with the makers of generators or electric locomotives without ever knowing who provided the copper without which they could not function.

This vulnerability extends also to real life. It was the fate of Thomas Bolton & Sons Ltd, having become a limited company in 1902, to be eventually taken over by the end-users of one of its main products, copper wire. In 1932, the family's holdings in the firm were largely bought out by Callender's Cable and Construction Co Ltd and British Insulated Cables. These merged in 1945 to form British Insulated Callender's Cables (BICC), and Boltons became a wholly owned subsidiary in 1961. Of the works originally established by the Boltons, Broad Street closed in 1911, Oakamoor was closed and demolished twenty years ago, Widnes has been considerably reduced in its importance, and the Sutton rolling mill ceased working in 1982. But the copper tradition is still continued at Froghall, where many forms of the metal are produced largely for the electrical industry.

To answer the second question, that is, why a copper works in the Churnet Valley in the first place, some reference to the earlier history of copper and brass working in the area is needed. The Oakamoor works which Thomas Bolton bought in 1852 had belonged to the famous Cheadle Brass Company, run by the Patten family and their partners. The Oakamoor copper and brass mill was built in 1792, but brass and wire making in the area go back much further to 1734, when Thomas Patten first built a brass works at Cheadle, three miles from Oakamoor, and a brass wire mill at Alton on the Churnet. The Patten firm itself goes back even further to 1719, when Thomas Patten built the first copper smelting works in Lancashire at Warrington. This takes the story back almost to the very beginning of the modern copper and brass industry in Britain, and it is appropriate perhaps on the occasion of a bi-centenary to begin by putting the story of Thomas Bolton & Sons Ltd in a wider historical context. For this reason the first chapter will tell briefly the not-widely-known history of the early brass and copper industry before the

Boltons' time, and in particular that of the Pattens, their predecessors at Oakamoor.

CHAPTER ONE

Before the Boltons 1734-1857

Copper was a late starter in British industrial history. While lead and tin had been mined and smelted since Roman times, and iron had been made throughout the Middle Ages, copper mining and smelting did not begin here until the reign of Queen Elizabeth I. Two monopoly companies were set up in 1568 to naturalise copper and brass-making in Britain with the help of German expertise and money. The most successful of these ventures was at Keswick in Cumberland where copper mining and smelting went on until the beginning of the Civil War in the 1640s. The Mines Royal and the Mineral and Battery Company, as the two ventures were called, proved to be a false start. No other domestic source of copper emerged after the Keswick smelter ceased. Those who tried to make brass with imported copper were finally forced out of business by pressure from the Dutch who had a monopoly of the copper and brass trade in Europe and did not want any English competition. By 1680, copper and brass making was virtually extinct.

The rise of the modern industry, which led to the establishment of Thomas Bolton & Sons, can be said to date from 1690. All aspects of the industry were present: mining, smelting, fabricating and manufacturing were all carried out during the revival of the industry. The two main reasons for the upsurge were the discovery of new sources of copper ore and the development of a revolutionary new way of smelting it.

As the Cornish tin mines sank deeper and deeper in the search for ore, by the 1680s the tinners began to find the lodes producing more and more copper. Cornwall was to become the basic source for ore till the 1840s, and it is true to say that up till that time most of the world's copper had come out of the ground in a few parishes in the county. The deep mines of Cornwall were later supplemented by the great open cast deposit on Anglesey known as Paris Mountain, by the Ecton mine in Staffordshire, and by Devon mines in the nineteenth century.

To smelt the sulphide ores from Cornwall an entirely new method was devised at Bristol in the 1680s, which remains the basis of copper smelting to this day. This new technology involved the use of a reverberatory furnace with coal as a fuel. In the traditional shaft or blast furnace the ore makes direct contact with the fuel, usually charcoal in the case of copper. Air is forced through the charge by bellows hence the name 'blast furnace'. In the reverberatory furnace, as it was developed by 1688, the fuel is separated from the ore and its heat is carried into the ore chamber by a powerful draught created by the design of the furnace and the use of a tall chimney. This unique British development allowed the use of coal for the first time in the smelting of metals. The new method made possible larger-scale smelting and made it cheaper than the traditional method which still prevailed in Europe. Bristol remained an important smelting centre in the early years, but the larger South Wales coalfield attracted more and more attention, and eventually became the main smelting district for domestic ores. Even after the British mines began to fade, much of the world's copper had to find its way to Neath and Swansea to be smelted. By the time the Boltons themselves became smelters in 1881, the centre of power was shifting particularly to the North American producers, who were soon to dominate the supply of copper especially for electrical purposes.

Brass making revived very quickly as English-made copper became available, although there were teething problems with using coal-smelted copper to make brass. Five major brass works were built between 1690 and 1703, near to their main markets in London and Bristol. Later came the Cheadle works and the brass works in North Wales producing for the developing Liverpool market. Birmingham saw its first brass-making works in 1740. Most of the copper smelter's output at first went into brass, which had most of the virtues of copper that were sought after at the time, and was cheaper into the bargain. Brass is tough enough to be hammered into all kinds of hollow wares, known as battery. It is ductile and was much used for wire for pins, for making wool-cards for the textile industry, for sieves of all kinds, and in the nineteenth century for weaving a fine mesh for paper-makers. C.H. Johnson and Sons Ltd, wire-weavers in Manchester for nearly two hundred years were among the customers of Thomas Bolton and Sons Ltd for brass wire to be used primarily for paper-maker's mesh. Brass has a fine colour and was much used then as now for ornamental purposes of all kinds, from buttons to coffin fittings. This was done largely by casting brass into moulds and later by stamping. Brass also resists corrosion. and so it was especially useful in the new steam engines which first appeared in the eighteenth century, where it was used for cocks, taps, valves and tubing of all kinds. Birmingham became the centre for the manufacture of hardwares from brass of both the decorative and functional kind.

Copper, by contrast with brass, had relatively few uses initially. It was used occasionally for coining farthings and halfpence. After 1730 it was exported to Asia in large quantities by the East India Company, where much of it ended up in bronze statues of the Buddha. Copper's resistance to corrosion found an important new outlet in the later eighteenth century, when it began to be used for sheathing the hulls of wooden sailing ships. The extensive trade in tropical waters carried out by the British merchant fleet meant that the timbers below the water-line were attacked by the rapacious toredo worm. Sheathing became standard practice for all ships, including those of the Royal Navy. Boltons were still making some copper sheathing in the 1880s, a reminder of the persistence of the wooden sailing ship.

Conductivity, the quality of most concern to the industry today, was known about in the eighteenth century. Following Benjamin Franklin's experiments with kites in thunder-storms, iron lightning conductors were fitted to St Pauls Cathedral and other public buildings. It was not till the development of current electricity that the special virtues of copper in conducting electrical currents were realised. A list of comparative conductivities was made by William Henley in 1774, which placed silvered copper after gold and brass. The quality of the copper at this time of course varied considerably in its level of impurities. The copper wire used by Faraday in his crucial experiments on electro-magnetic induction between 1821 and 1831 has been recently analysed. The conclusion was reached that had the conductivity been only a little worse, Faraday's work would have been affected. Concern over the conductivity of copper first arose with the development of the electric telegraph, but this story will be dealt with later, as will the contribution of the Boltons in developing high-conductivity copper.

The last aspect of the new copper and brass industry worth mentioning here is finance. Most of the money went into the smelting works and the stocks of ore required to keep them running. The largest works had thirty-three furnaces, and needed by the standards of the time a substantial capital expenditure of four or five thousand pounds. It was the usual practice to keep at least three months supply of copper ore on hand at the works, which also tied up a lot of working capital, possibly as much as £10,000. Only relatively wealthy merchants had this kind of ready cash and the industry was financed almost exclusively by business men from London, Bristol, and later the Liverpool area, the main centres of commercial activity in the eighteenth century. A handful of large smelting companies dominated the market during the first century of the industry's existence, and they used their influence ruthlessly to exclude newcomers and fix prices to their own advantage, practices which will be quite familiar to twentieth-century readers. Small manufacturers and users of copper and brass, particularly in Birmingham were the losers. The Bolton family began their business in Birmingham just at the time when the ring of 'associated smelters', or 'brazen sovereigns' as one contemporary called them, were forcing on Birmingham an increase in the price of copper and brass.

Thomas Patten and the Cheadle Brass Company

Thomas Bolton and Sons' predecessors in Staffordshire, the Cheadle Brass Company, began during the rise of the industry outlined above. The founder of the Patten industrial dynasty, which survived for four generations, was Thomas Patten, junior, 1690-1772. He came from a mercantile Lancashire family, then living in Warrington, a small town on the Mersey. His father was an active trader along the Mersey and was responsible for making the river navigable for ships from Runcorn to Warrington in the 1690s. This made possible the building of the first copper smelting works in Lancashire on Bank Quay in Warrington in 1719. Copper ore and coal could now be bought with ease down the river to the works. The works were not large: by 1754 they still had only twelve reverberatory furnaces. Originally Warrington was conceived of as part of an ambitious partnership which embraced copper mines in Cornwall, and iron works on the Severn as well as the smelter. Thomas Patten's partner was William Wood, who had inspired the scheme, and attempted to capitalise on it during the South Sea Bubble of 1720. However, he lost his money in the crash, and eventually sold out his share in 1725. Thomas Patten's father died the following year, and he inherited the family businesses. A contemporary in the copper trade noted in 1728 that the Warrington works 'do but little'. It was clear by that time that the best profits were coming from brass-making, not just from smelting. So it is natural to find Thomas Patten in 1734 setting up a brass-works and a brass-wire mill to make use of his copper from Warrington.

The site was more surprising, since the works were well away from Warrington, in the north-east corner of Staffordshire, near Cheadle and at Alton.* The copper had to come overland by packhorse, an expensive method of carriage compared to water transport. But the new markets for brass in Birmingham and the Black Country were not far away, and much of the wire went to the pin makers in the south of England, at Gloucester for example.

Two brass-works were set up on what was then known as Spout Farm, now the suburb of Cheadle called Brookhouses. The first, 'Old Spout', was originally built to provide brass for the wire works. The second, 'New Spout', was privately run by Patten and made ingot brass for the Midland hardware market.

---

* Note: The widespread story that the brass and copper industry came to the Cheadle area due to the rich supplies of copper ore from the Ecton Copper mines, about eleven miles away can be completely discounted. When Thomas Patten set up his brass works at Cheadle and the brass-wire mill at Alton in 1734, the Ecton mines were only being worked on a small scale and the main rich body of ore had not been discovered. The ore was sent to Denby, near Derby, for smelting and there is no evidence that any of the copper metal came to Cheadle at that period.

It was only during the 1760s that some of the Ecton ore went to Patten's copper works at Warrington. The copper works built at Cheadle in 1788 to take advantage of the increased output from Ecton were a disappointment as the owner of the Ecton mine, the Duke of Devonshire, built his own smelting works at Whiston two years later.

The brass industry, and only later the copper industry, came to Cheadle because of the combination of coal supplies, water power and a location already part of the way towards the metal's ultimate market in the Birmingham area.

The methods used in the 'melting-houses' at Cheadle for making brass were the traditional ones used for centuries in Europe. The alloy of copper and zinc was made by mixing copper with calamine and charcoal. The copper was granulated, and was known as 'copper shot', an English innovation dating from 1723. Calamine, or zinc carbonate, was found in quantities in the nearby lead mines of Derbyshire. It was calcined, or roasted, to turn it to zinc oxide before being ground up by a large stone wheel. The three ingredients were then put into clay crucibles or 'pots' about one foot high, in the proportions of one part copper to two parts calamine, by weight. The crucible was sealed and put into the furnace. Eight small pots were arranged in a circle around one 'king pot'. Many of these 'pots' can still be seen in Cheadle garden walls with their pointed bottoms broken off, serving as drainage pipes. Heat was applied from the coal fire below, whose fuel was largely supplied from the Woodhead seam mined only a mile from the works. The zinc vapourised and penetrated the hot copper to make brass.

Every twelve hours, the Cheadle brass-makers would pour the contents of the small pots into the king pot. Any slag was cleared from the top, and the brass was ready for the next stage. That intended for the wire mill was made from 'best copper' and was poured into granite moulds to make a 'slip' 4½in wide, 4ft 6in long and ¼in thick. Ingot brass, made with 'common copper', was poured into iron moulds to make slabs ready for remelting by the brassfounders of the Black Country, Birmingham and other places.

The work was arduous and required skill and judgement if the charge was not to be lost. All the brass works which began the new industry employed imported Dutch or German brass melters, and Thomas Patten's was no exception. He brought John Essor from Germany at a cost of ten guineas in travelling expenses, and two guineas in fees to the agent who persuaded him to come. Essor changed his name to Keys, and his descendants acted as chief melters at both brass works throughout their history, later coming to own them themselves.

The calamine or cementation method of making brass was the standard method at the time of the Boltons' beginnings in the Birmingham area, and was brought to its greatest perfection by the Cheadle Brass Company. Its replacement by the direct method of making brass with zinc metal or 'spelter' marked the end of their hegemony in the trade. Brass-making by the calamine method at Cheadle ceased in 1830, and hardly any brass was being made in that way anywhere by 1850.

The method used to make brass wire at Patten's Alton mill survived rather longer. The same basic principle applied to the manufacture of all wire at that time. The metal was rolled into thin sheets, which were slit into square-sectioned rods. These were then drawn down through dies into round wire.

The wire mill was situated at the northern end of the bridge over the River Churnet at Alton. The 'slips' arrived there by horse-back and were taken first to the rolling mill. Here water-powered rolls, 12in diameter and width, and made of highly polished cast iron, reduced the plates from $\frac{1}{4}$in thickness to $\frac{1}{16}$in. They were then taken to the slitters. These were sets of iron discs with edges so sharp that, as a Swedish visitor observed, a piece of paper 'will be cut as neatly as with a pair of scissors'. Patten's slitters consisted of two sets of discs, each 12in diameter. The second set were arranged behind the first, with the discs fitting into the spaces between those of the one in front. In this way, the water-powered slitters reduced a 'slip' $4\frac{1}{2}$in wide to twenty-six square rods, about $\frac{1}{16}$in by $\frac{1}{16}$in. These were then taken to the wire-mill proper. 'Ripping' or 'rumpling' came first, where the square rod was drawn seven times through one die to reduce it to a round one. This was passed to the fine wire mill where it was drawn down further through seven dies by a water-powered drawing block, or capstan, made of wood. The wire was covered with lard or tallow to lubricate its passage through the dies. Rolling and drawing work hardens the brass, so that it has to be annealed by heating till red hot at each stage.

The length of the final piece of wire was obviously dictated by the length of the original 'slip'. For brass wire, this was not critical, and the technology of brass wire-making remained fundamentally the same even into the twentieth century. In 1906, E.J. Bolton described rolling, slitting and drawing as still being the main processes in making it in England, though not in the USA. The only difference was that by then the strips of brass were made by flattening brass rods and not by casting brass sheets. With copper wire, however, it was a different matter. With the coming of the electric telegraph, and later the telephone and long distance power transmission, having only short lengths of conductor was a serious limitation. Frequent joints were needed and this led to problems in manufacture and usage. How these problems were progressively tackled and overcome forms part of the story to come.

The Staffordshire works of Thomas Patten and Company were built mainly to satisfy an inland market for brass wire and ingot brass. But a large new outlet for copper was growing rapidly much nearer to Warrington, in the port of Liverpool. This was the ever-growing demand for trade goods for the slave trade. In 1737, thirty-three ships had left Liverpool for the West African coast. By 1753, the number was seventy-two, and by 1771, it had reached 107. Items made of brass and copper figured largely in this trade. Shallow brass battery dishes called 'neptunes' were popular, as were brass and copper rods, and 'manillas', horse-shoe shaped rods with flattened ends, used for ornament and also as a form of currency. There was also a substantial trade with the slave plantations for copper stills for sugar boiling and other purposes. Patten was not slow to respond to this demand, and began to extend his activities to the Greenfield valley, near Holywell in North Wales. He had already acquired a site there for industrial development by 1743. By 1753 he had a copper mill making 'Guinea rods' for the African trade, and for rolling copper for copper battery. Later a brass wire mill was added.

In 1767, Patten's Greenfield interests, and the Cheadle brass ingot house, were brought together in a separate partnership called the

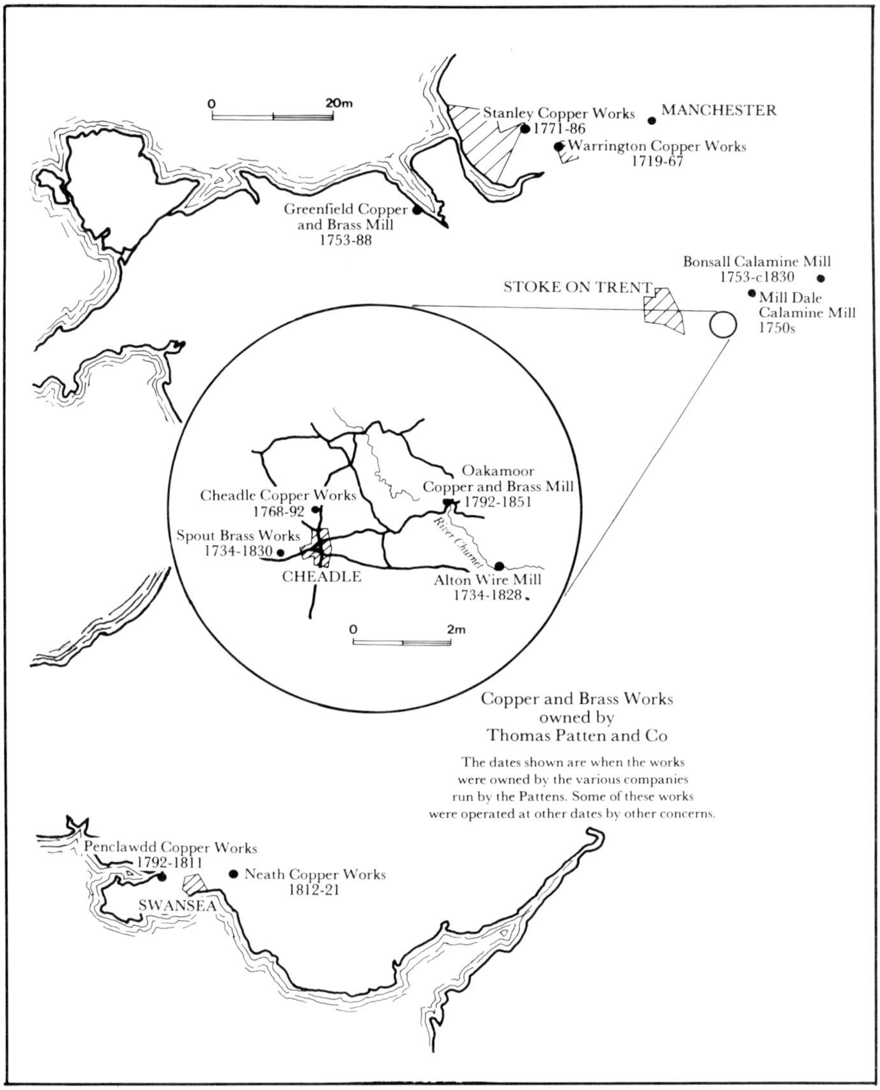

Warrington Copper and Brass Company. This also included copper smelting in its activities, but the Warrington smelter was now leased out and for the first time copper was smelted at Cheadle itself. A new copper smelting works was built near the Woodhead colliery in 1768, to smelt the ores now coming in large quantities from the Duke of Devonshire's copper mine at Ecton. Copper still came from Warrington to Cheadle, but it was gradually being replaced by the local Cheadle Copper Works supply, which was also taken to the Greenfield works.

Patten's empire was further extended in 1771, when he built the Stanley copper smelting works on the outskirts of St Helens, in the township of Ashton-in-Makerfield. This was built on the Sankey Brook canal, and the aim was to ship in ore from the new open-cast Mona mine on Anglesey. But his operations were not large enough to withstand the pressure from the 'Copper King', Thomas Williams, who gained for a few years towards the end of the eighteenth century almost complete personal domination over the entire copper and brass trade in Britain, through his control of the Anglesey copper mines. Patten eventually sold his interests in the Greenfield valley and the Stanley copper works to the Welshman, and in 1788, consolidated around Cheadle and Alton. A tin-plate works on the River Churnet at Oakamoor was bought in 1790 and converted into a rolling and slitting mill for brass and copper to supply the Alton wire mill, and also to draw thick wire for making 'Guinea rods'. Within a few years, he had acquired a new smelting works at Neath Abbey in South Wales, and by 1800 another at Penclawdd, on the Gower peninsula. The Woodhead smelter was closed, and operations removed to the brass works site.

It was on these works that the reputation of Cheadle Brass was based. It was regarded by expert contemporaries as 'probably the best in Europe', and therefore in the world. The works produced up to 600 tons of brass a year in the early nineteenth century, from thirty-six furnaces in four melting houses. Three-quarters of this went into wire or sheet brass for the hardware trade. The rest was made into brass ingots weighing 6 to 7lb, for use by brass founders. Most of the firm's output, perhaps 350 tons a year of wire, ingots and 'latten' or sheet brass, went to Birmingham. Some of the inferior 'ash metal' marked YY went to Wolverhampton and Sheffield, and the latter town also took some of the top grade 'BB' brass, made with best copper. The other grades were 'BC', made with common copper and more zinc, and 'AM' which included some lead.

This was the 'golden age' of the Cheadle Brass Company. By 1795 the stock was worth a total of nearly £60,000. Within ten years its capital

The obverse and reverse of the one penny copper token issued by the Cheadle Copper and Brass Company in 1812.

Interior of the strip drawing mill at Oakamoor, photographed in 1946. This was one of the Cheadle Copper and Brass Company's early buildings, erected in 1804. The arches were often likened to those in a church or chapel.

value was reckoned to be over £100,000. However, a lot of this was due to borrowings during the inflation of the Napoleonic War period. In the depression that followed the end of the war, the company was forced to retrench. In 1821 all salaries and wages were cut, and the Neath Abbey works wound up. The man sent to do this was William Keates the son of the brass works manager. Shortly after the closure of Neath Abbey, Keates left the company and began a new career for himself which will cause him to be mentioned again at a later stage.

In the 1820s, the Cheadle Brass Company also began to be affected by something more fundamental than a trade depression. The basis of their reputation, the making of high quality calamine brass, was being undermined by changes in technology, and the steadily increasing manufacture of brass by the direct method using metallic zinc or 'spelter'. Some high quality brass had always been made in this way, as the calamine method would not allow the manufacture of a brass with a very high copper content. The factor limiting the general use of the direct method was the expense of metallic zinc, which could be as much as £280 a ton in the eighteenth century, compared with £24 for the equivalent zinc content of calamine. Originally imported from Germany and the Far East, spelter began to be made in Britain commercially under a patent taken out by William Champion of Bristol in 1738. But the first attempt to make brass on a large scale using home-made zinc had to wait till 1781, when a former manager of William Champion's copper works James Emerson took out a patent. His brass, made at his own works at Hanham near Bristol, was regarded at the time as the 'purest in the world' and was in great demand in Birmingham and elsewhere. However, Emerson went bankrupt in 1803, and the development of brass making by the direct method was held back. Other manufacturers began to make it, but for some time, they could not match the quality of the best calamine brass such as that made at Cheadle. This was possible due to their not using spelter as pure as that made by Emerson. Birmingham gilt button-makers also said that calamine brass received the gold better than spelter-made brass. Calamine brass certainly kept its colour better than zinc brass. But by the 1820s these differences in quality were diminishing and moreover zinc brass was becoming cheaper than the old calamine variety. Because it required less extensive plant to make, zinc brass could be made by the brass-user himself, which undermined the market for Cheadle's ingot brass, sheet and wire. The partners became aware of the Birmingham competition in 1819, noting with alarm:

◀◀ *Doorway of the Cheadle Company's wire mill at Oakamoor, photographed shortly before demolition in 1963. The inscription reads* T[homas] P[atten] & C[ompany] 1827. *This building was later the tube mill (or 'solid shop').*

◀ *Clock and plaque dated 1829 on the main offices at Oakamoor.*

> a number of persons in Birmingham manufacturing wire which they can afford to sell very low by reason of the expense of their establishments being very trifling.

Closing the Cheadle Brass works was considered then, but not carried out. The Alton wire mill in fact was the first to go. In 1828 it was decided to concentrate the wire production at the newer works at Oakamoor, which had up till then been mainly used for rolling and slitting. A new wire mill was built there.

This was not quite the end of Alton. The redundant manager, John Brindley, attempted to keep it going on his own account. But he went bankrupt in April 1834, and the Alton Mill, 'ex Cheadle Brass Company', was to let. The machinery was advertised for sale in August. Although it never functioned as a wire mill again, Alton continued to be used industrially, and by February 1836, it had been taken over by a paper-making partnership. Some remnant of the original 1736 building still stands, and there is a datestone in the wall inscribed 'T.P. & Co 1736'.

In 1830 it was the turn of Cheadle itself to be closed down. Brass-making was transferred to Oakamoor. In October all redundant 'hands', including the manager John Keates, were given two months notice. He appealed in vain for some redress from the company.

By the end of 1830 all brass-making by the company at Cheadle itself had ceased. But the works were kept going to some extent by James Keys, a descendant of the original German artisan brought over by Thomas Patten in 1735. He had been the company's copper refiner, and since 1829, no doubt having seen the writing on the wall as far as Cheadle was concerned, had also become the manager of the nearby Whiston copper works, where the ore from the Ecton mine was smelted. James, and his son John after him continued to lease the Cheadle site until J.W. Patten put it up for sale in 1856. John Keys then bought it outright for £2,150. At that time it consisted of the old disused calamine brass works, and a copper and spelter smelting works, which was the business Keys had been carrying on there. James Keys had already become the owner of the Whiston works in 1847, so they were now substantial figures in their own right in the industry. Whiston closed finally in 1890. The Cheadle works closed at the same time and James Keys, son of John, asked the Boltons to take the men into Oakamoor or Widnes. The last of the family, John Keys, died in 1907. The last landmark of the works, the giant copper furnace chimney, was demolished in 1907. Since then the site has been derelict. Now surrounded by buildings, it remains as a mute relic of a long and illustrious industrial tradition.

The closure of Alton and the Cheadle works meant that all the company's activities were concentrated at Oakamoor. There is no detailed information about the company during the 1830s, but by the next decade there had been little or no change in the product line. In 1844, Oakamoor was still producing sheet brass and copper, and brass wire of various gauges. Twenty-seven men were employed on the rolls and their annealing furnaces, nine on slitting and winding. The square rods from the slitters were wound round a wooden spindle before being annealed again. The spindle was turned by a crank, and this was often a job for young boys in the works. Ten worked on the 'rumpling', or first drawing of the square rods, and twenty-eight in the wire drawing shop. The bill for a fortnight's wages for the whole labour force of 114 in October 1844 was £151 3s 2d, an average wage of 13s 3d or 66p per week. The head workmen got more, the less skilled less than this. A glimpse of the working conditions there can be had from three statements made to the Commissioners on the Employment of Children in 1842. These are worth reproducing in their entirety, to give their full flavour.

CHEADLE BRASS AND COPPER FOUNDRY, OAKAMOOR

No 35 Joseph Jeffries, aged 17. Examined 23 March:

No 35 I have been here to work five years, as a wire-drawer. I went to day-school at Oakamoor before that for two years. I went to Sunday-school there too; I can read pretty well, and write. There are two lads under 18 working in the same room with me as wire-drawers, both of them can read, and one can write. I come to work at six o'clock and go home at six; two hours are allowed out of this for meals, we always go when the bell rings and come in too; I work by day wages for master, not for men, all the rest do the same; there are no bargain wages of any kind; I get 9s a week; I am not bound an apprentice, none of the rest are either. I never met with any accident in the machinery, I never saw anybody else meet with accidents at the work I am at; about twelve months ago, a boy had his arm taken by the pinion of the wheel, he nearly lost his life; but by great attention to him, he recovered and is now the post-boy for the works and parish; it was his own fault, as he was staring about when he ought to have been greasing the wheel. I have never no sickness or ill health; I like the work very well, it is easy enough. I live with my sister, as I have no father or mother; she is married, she has nothing else but what my wages bring. I have enough to eat and drink.

No 36 William Finney, aged 16. Examined 23 March:

No 36 I work in the smithy of the Cheadle Companies' Brass Works; my brother works there with me, he is younger than me. I went to day-school before I came to work, and to Sunday-school; I have worked here a year and a half; I can read and write. I come to work at six o'clock with all the other people, when the bell rings, go to breakfast at eight, return at half-past eight, go to dinner at 12, return at one, go to tea at four, come back at half past, and leave work at six. I do this regularly six days in the week, my wages are 6s; my brother's 6s; I have another brother a wire-drawer getting 4s; none of us are apprentices; the overseers and workmen are all very good to us upon the works, they are never known to ill-use us. All of us work by day wage, we are paid every other Thursday in hard cash; both myself and brothers are in good health; we never met with any accident, one of my brothers did by getting his hand caught in a pulley; he recovered his hand again, as well as ever. We have nothing to complain of from anybody, and are very happy and comfortable.

No 37 Thomas Wilson, aged 74. Examined 23 March:

No 37 I have worked for this company 63 years; I think I am the oldest servant of the company but I have two brothers who have just given over wire-drawing, one is 82, the other 85; they have worked here ever since they were able, they began so long before me that I do not remember their beginning. There are twelve or fifteen men now upon the premises as old and older than myself, whose united ages amount to almost a thousand years. In my time a vast many men and boys have been brought up to the business; our ages are sufficient proofs that there is nothing unwholesome in the nature of it. I have seen a few accidents with the machinery, but these accidents in a general way are resulting from carelessness. I have never known any lives lost; our hours of work are very regular, and average twelve hours a day; if there should be any extra demand we make seven days a week or fourteen hours a day deducting two hours for meals. We all come at the same hour and go home at the same, there is no difference with any of us, when the bell rings it is a sufficient notice. The boys in the works conduct themselves very well for anything I know to the contrary; they are all intelligent boys and attend at the desire of Mr Wrag, the church or chapel, and Sunday-schools; I believe they are better educated than they used to be, they have a better chance than they had in my time; my wages are a pound a week, I am paid like all the rest on every other Thursday, always in cash. Our workmen are well-conducted, quiet, orderly men, they behave very well to the boys upon the works, and to their children at home; there are no punishments of any kind; if a man misconducts himself, our masters will not employ him; they will not keep drunkards upon the premises to be bad examples to the young ones. I think we ought to be the thankfullest of men, for we have the best masters in England, God bless 'em.

I cannot write, I never went to school when a boy; I can read the Scriptures Thank the Lord — I would have every child do that much.

The commissioners noted that the works were very extensive, and the rooms very spacious, clean, light and well-ventilated. Although the pay was not high even by the standard of the time, there were evident compensations in the conditions of work. Life in a copper smelting works, for example, would not have been so healthy, although the pay was higher. Ironically, it was their own employer, John Wilson Patten, MP for North Lancashire, who had brought about the Royal Commission. He had opposed Lord Ashley's original Bill to limit the hours of employment of women and children, and carried by a majority of one his own motion for a Royal Commission to enquire further into the question.

The partners attempted to widen the scope of the works in the 1840s. In 1845, the number of casters, or brass makers was increased from four to ten, working on three furnaces. In the following year, the company began, rather belatedly, to make brass and copper tubes, This was a product which had for some time been a Birmingham speciality. The market had increased enormously as a result of the new railway mania, and the demand for boiler tubes for steam locomotives. John Collier was employed as chief tube-maker, with six men and six boys under him.

But for Oakamoor, an era was closing. Production had been in decline since 1844. 380 tons of copper had been bought in that year, but this fell to 288 in 1848. In the final three years of Patten ownership copper purchases fell further from 214 tons in 1849, to 127 in 1850, and only 20 in the final year 1851. By Christmas most of the men had been paid off, the last few going in the New Year.

Why did this celebrated firm decide to close down, just at the point when a vast new market for copper wire was opening up with the advent of the electric telegraph? The answer seems to lie in the social attitudes which the Pattens had developed over the years. By the end of the eighteenth century, Thomas Patten II had a house in Bath, and clearly had the outlook of a gentrified businessman of that time. Technical change was something that could be left to men of artisan or managerial status, while they were left to pursue higher aspirations. This is reflected in the upbringing of his grandson, J.W. Patten, the last member of the family to engage in the industry. He received a gentleman's education at Eton, and then Magdalen College, Oxford. This was completed by making the 'Grand Tour' of Europe. He was elected for Parliament in 1830, and again in 1832 with the help of Lord Derby, whose son he had met at Oxford. Patten served continuously for the North Lancashire division until 1874, filling governmental posts under Lord Derby on the way. In 1874 he was made Baron Winmarleigh by Disraeli. Patten's interests obviously lay primarily on the floor of the House of Commons and not on that of the Oakamoor wire-drawing shop. A parliamentary career came before the development of his industrial interests. So he was prepared to let Oakamoor go to the Boltons in 1852, relatively cheaply.

It is not surprising that J.W. Patten's father did not consider that a technical or scientific education was appropriate for his son, and sent him instead through the traditional establishment institutions. This was not universal in the industry. John Henry Vivian, the son of the Swansea copper smelter John Vivian, was sent to learn languages at Marburg and mining and metallurgy at Freiburg. His 'Grand Tour' was a scientific one through the mining districts of Germany. A scientific education was however considered suitable for Thomas Patten's employees. William Keates, who has already been mentioned, was sent at the firm's expense in 1819 to study chemistry in London. It was no accident that both John Vivian and William Keates were able to play an important part in the industry of their time, where John Wilson Patten was not. As the century progressed, more and more scientific knowledge became necessary in industry. The whole electrical industry sprang literally out of Faraday's experiments in the laboratories of the Royal Institution. Only those who could follow these developments and grasp their significance would be able to take advantage of them commercially. The Pattens remained relatively indifferent to these matters and their industrial activities suffered accordingly.

This failure by the Pattens was symptomatic of a more general problem. The curious paradox which caused the first industrial nation to

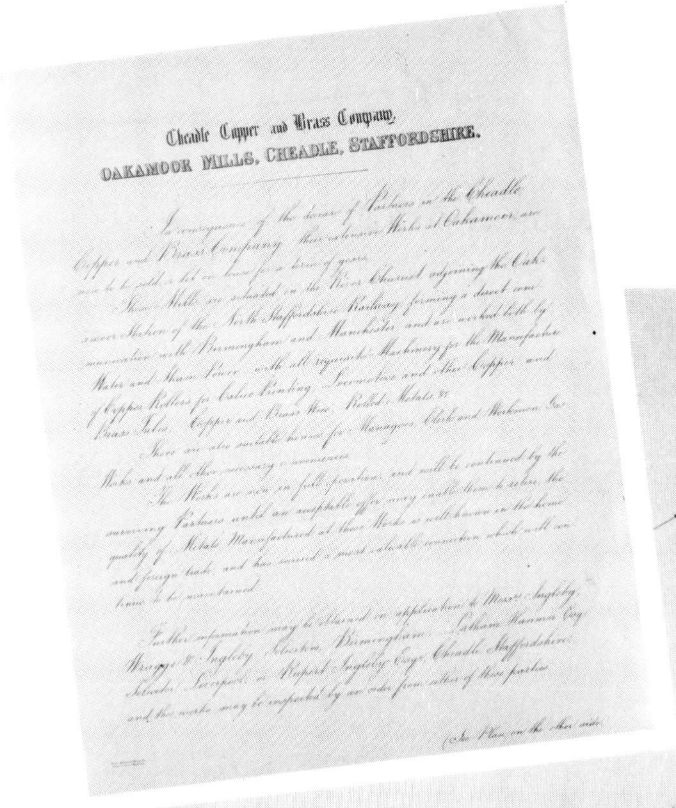

Cheadle Copper and Brass Company,
OAKAMOOR MILLS, CHEADLE, STAFFORDSHIRE.

In consequence of the decease of Partners in the Cheadle Copper and Brass Company their extensive Works at Oakamoor are to be sold or let on lease for a term of years.

These Mills are situated on the River Churnet adjoining the Oakamoor Station of the North Staffordshire Railway forming a direct communication with Birmingham and Manchester and are worked both by Water and Steam Power with all requisite Machinery for the Manufacture of Copper Rollers for Calico Printing, Locomotive and other Copper and Brass Tubes, Copper and Brass Wire, Rolled Metals &c.

There are also suitable houses for Managers, Clerk and Workmen for the Works and all other necessary conveniences.

The Works are now in full operation and will be continued by the surviving Partners until an acceptable offer may enable them to retire. The quality of Metals Manufactured at these Works is well known in the home and foreign trade, and has secured a most valuable connection which will continue to be maintained.

Further information may be obtained on application to Messrs Ingleby, Wragge & Ingleby, Solicitors, Birmingham; Latham Hanmer Esq. Solicitor, Liverpool; or Rupert Ingleby Esq. Cheadle, Staffordshire, and the works may be inspected by an order from either of these parties.

(See Plan on the other side)

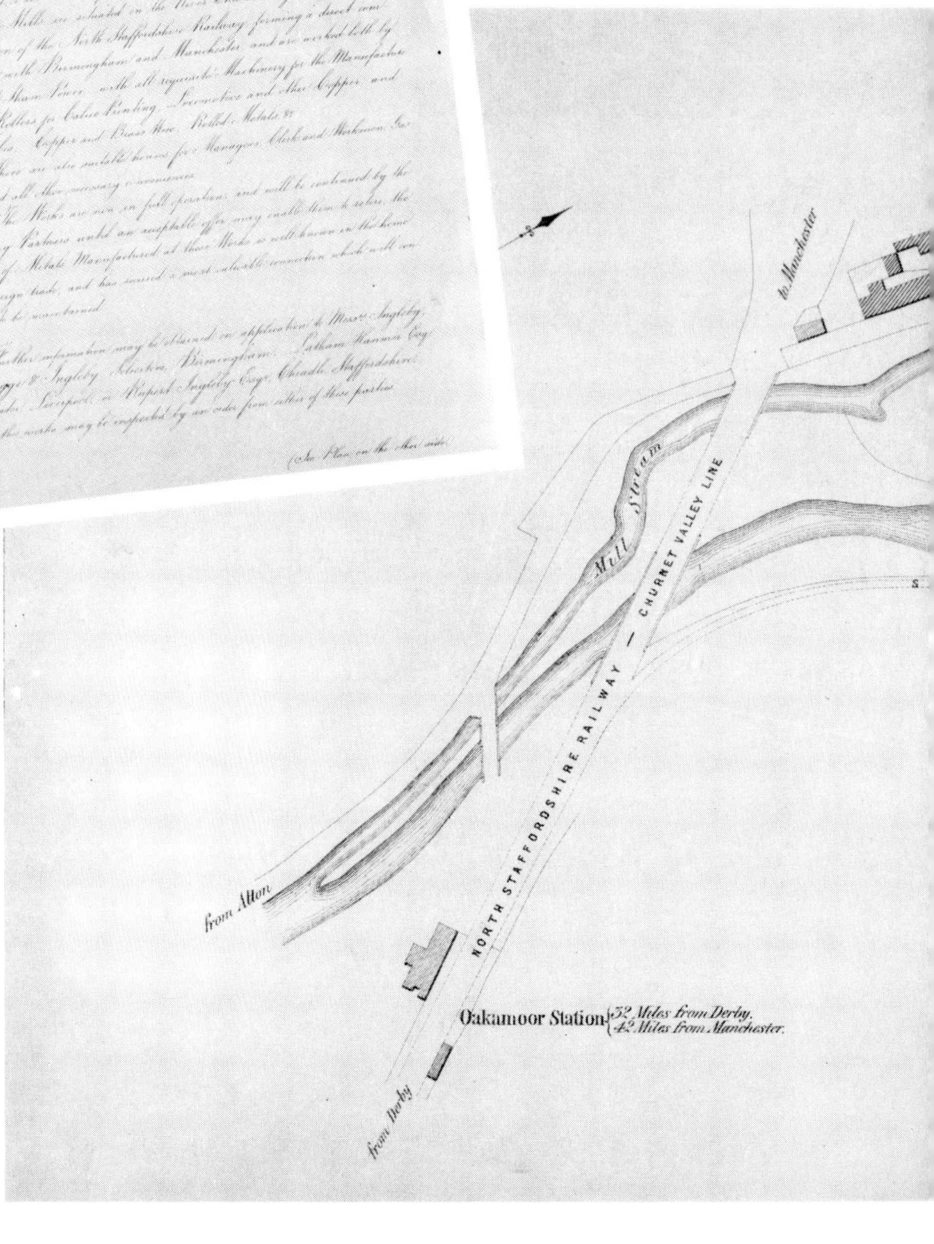

Sale notice and plan of the Cheadle Copper and Brass Company's Oakamoor works in 1851. No acceptable offers resulted from this sale.

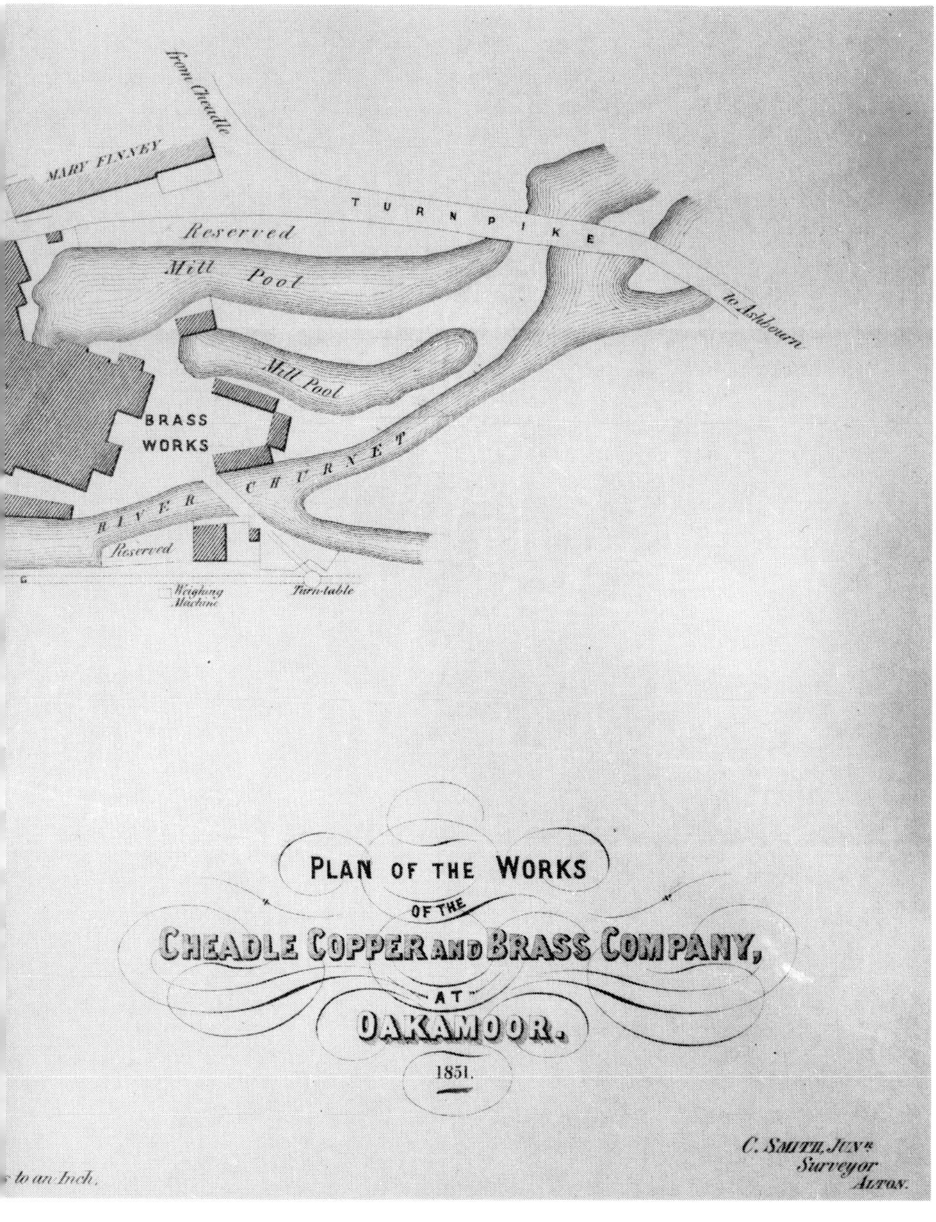

educate its leaders in a spirit quite alien to the growing needs of business and technology was one of the reasons for the loss of Britain's industrial lead in the later nineteenth century. The Boltons however had a more enlightened attitude. A few years after taking over Oakamoor, Alfred Bolton opened a reading room at the works. He gave the first lecture there, entitled 'The Pleasures, Objects and Advantages of Science'. It was in this spirit that the first Thomas Bolton had educated his sons, and the tradition was to continue through the next generations.

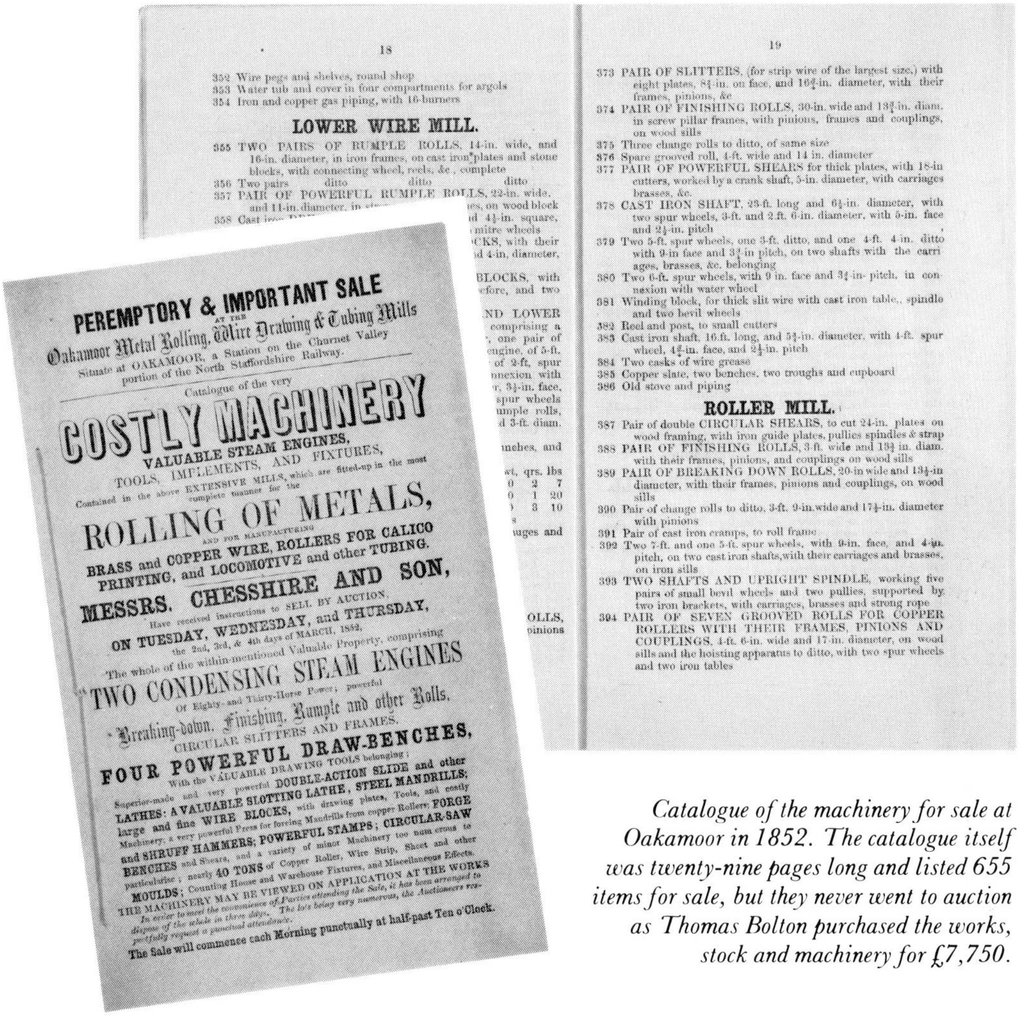

PEREMPTORY & IMPORTANT SALE
AT THE
Oakamoor Metal Rolling, Wire Drawing & Tubing Mills
Situate at OAKAMOOR, a Station on the Churnet Valley portion of the North Staffordshire Railway.

Catalogue of the very
COSTLY MACHINERY
VALUABLE STEAM ENGINES,
TOOLS, IMPLEMENTS, AND FIXTURES,
Contained in the above EXTENSIVE MILLS, which are fitted-up in the most complete manner for the
ROLLING OF METALS,
AND FOR MANUFACTURING
BRASS and COPPER WIRE, ROLLERS FOR CALICO PRINTING, and LOCOMOTIVE and other TUBING.

MESSRS. CHESSHIRE AND SON,
Have received instructions to SELL BY AUCTION,
ON TUESDAY, WEDNESDAY, and THURSDAY,
the 2nd, 3rd, & 4th days of MARCH, 1852,
The whole of the within-mentioned Valuable Property, comprising
TWO CONDENSING STEAM ENGINES
Of Eighty- and Thirty-Horse Power; powerful
Breaking-down, Finishing, Rumple and other Rolls,
CIRCULAR SLITTERS AND FRAMES,
FOUR POWERFUL DRAW-BENCHES,
With the VALUABLE DRAWING TOOLS belonging;
Superior-made and very powerful DOUBLE-ACTION SLIDE and other LATHES: A VALUABLE SLOTTING LATHE, STEEL MANDRILLS; large and fine WIRE BLOCKS, with drawing plates, Tools, and costly Machinery; a very powerful Press for forcing Mandrills from copper Rollers; FORGE and SHRUFF HAMMERS; POWERFUL STAMPS; CIRCULAR-SAW BENCHES and Shears, and a variety of minor Machinery too numerous to particularise; nearly 40 TONS of Copper Roller, Wire Strip, Sheet and other MOULDS; Counting House and Warehouse Fixtures, and Miscellaneous Effects.
THE MACHINERY MAY BE VIEWED ON APPLICATION AT THE WORKS
In order to meet the convenience of Parties attending the Sale, it has been arranged to dispose of the whole in three days. The lots being very numerous, the Auctioneers respectfully request a punctual attendance.
The Sale will commence each Morning punctually at half-past Ten o'Clock.

18

352 Wire pegs and shelves, round shop
353 Water tub and cover in four compartments for argols
354 Iron and copper gas piping, with 16-burners

LOWER WIRE MILL.

355 TWO PAIRS OF RUMPLE ROLLS, 14-in. wide, and 16-in. diameter, in iron frames, on cast iron plates and stone blocks, with connecting wheel, reels, &c., complete
356 Two pairs ditto ditto ditto
357 PAIR OF POWERFUL RUMPLE ROLLS, 22-in. wide,

19

373 PAIR OF SLITTERS, (for strip wire of the largest size,) with eight plates, 8¾-in. on face, and 16⅝-in. diameter, with their frames, pinions, &c
374 PAIR OF FINISHING ROLLS, 30-in. wide and 13¾-in. diam. in screw pillar frames, with pinions, frames and couplings, on wood sills
375 Three change rolls to ditto, of same size
376 Spare grooved roll, 4-ft. wide and 14 in. diameter
377 PAIR OF POWERFUL SHEARS for thick plates, with 18-in cutters, worked by a crank shaft, 5-in. diameter, with carriages brasses, &c.
378 CAST IRON SHAFT, 23-ft. long and 6½-in. diameter, with two spur wheels, 3-ft. and 2.ft. 6-in. diameter, with 5-in. face and 2½-in. pitch
379 Two 5-ft. spur wheels, one 3-ft. ditto, and one 4-ft. 4-in. ditto with 9-in face and 3¼-in pitch, on two shafts with the carriages, brasses, &c. belonging
380 Two 6-ft. spur wheels, with 9 in. face and 3¼-in- pitch, in connexion with water wheel
381 Winding block, for thick slit wire with cast iron table,, spindle and two bevil wheels
382 Reel and post, to small cutters
383 Cast iron shaft, 16.ft. long, and 5¼-in. diameter, with 4-ft. spur wheel, 4⅝-in. face, and 2½-in. pitch
384 Two casks of wire grease
385 Copper slate, two benches, two troughs and cupboard
386 Old stove and piping

ROLLER MILL.

387 Pair of double CIRCULAR SHEARS, to cut 24-in. plates on wood framing, with iron guide plates, pullies spindles & strap
388 PAIR OF FINISHING ROLLS, 3-ft. wide and 13¾ in. diam. with their frames, pinions, and couplings on wood sills
389 PAIR OF BREAKING DOWN ROLLS, 20-in wide and 13½-in diameter, with their frames, pinions and couplings, on wood sills
390 Pair of change rolls to ditto, 3-ft. 9-in. wide and 17½-in. diameter with pinions
391 Pair of cast iron cramps, to roll frame
392 Two 7-ft. and one 5-ft. spur wheels, with 9-in. face, and 4-in. pitch, on two cast iron shafts, with their carriages and brasses, on iron sills
393 TWO SHAFTS AND UPRIGHT SPINDLE, working five pairs of small bevil wheels and two pullies, supported by two iron brackets, with carriages, brasses and strong rope
394 PAIR OF SEVEN GROOVED ROLLS FOR COPPER ROLLERS WITH THEIR FRAMES, PINIONS AND COUPLINGS, 4-ft. 6-in. wide and 17-in. diameter, on wood sills and the hoisting apparatus to ditto, with two spur wheels and two iron tables

Catalogue of the machinery for sale at Oakamoor in 1852. The catalogue itself was twenty-nine pages long and listed 655 items for sale, but they never went to auction as Thomas Bolton purchased the works, stock and machinery for £7,750.

CHAPTER TWO

The Beginnings at Birmingham 1783-1853

Thomas Patten had begun his family's industrial career with a proverbial silver spoon provided by his father's merchant wealth. The Bolton story begins very differently. Birmingham and its surrounding Black Country towns were built up from innumerable small businesses working with little capital, often employing only a few people, frequently in small, cramped workshops. The celebrated 'Soho Manufactory' of Matthew Boulton, built as it was on open heathland outside the town, was not typical of the Birmingham of its time. Few of the crowned heads of Europe who visited Boulton's 'palace of industry' ventured into the dark courts and narrow streets in the town centre where the majority of the town's craftsmen lived and worked. It was in such surroundings that the Bolton family business began.

The Boltons did not originate in the town. They moved to it in the 1780s from Walsall, a Black Country town ten miles to the north. Like its neighbours, Willenhall, Wednesfield and Wolverhampton, Walsall specialised in certain metal trades. Originally, and still today famous for saddlery and leather goods, Walsall had developed a considerable trade in horse harness and trappings made from iron, steel and brass. Then in the eighteenth century, bucklemaking appeared as its second speciality. Shoe buckles were an essential part of the fashion of the time, and Walsall buckles were in demand all over Europe.

The first of the family about whom there is any detailed knowledge is Samuel Bolton, grandfather of the Thomas who gave the firm his name. Samuel was established for many years in New Street, Walsall as a whitesmith, that is, a maker of edge tools such as scythes, reaping hooks, axes and hatchets. Married in 1748, his wife Mary died in 1778 and was buried in Walsall. He had two sons, Richard and John, and when he died in 1781, he willed his business in New Street to Richard, and £70 to the younger boy, John, to be paid to him from the proceeds of the sale of his personal estate. His wife having died, another Walsall whitesmith, Thomas Dudley, and one Joseph Longster were made executors. There is no mention of his brother Thomas in his will. Thomas had also lived in Walsall and his wife Dorothy died there in 1772.

The move to Birmingham came after Samuel Bolton's death. The town was then growing very rapidly. In 1770 it was bigger than Manchester,

The Bolton Family

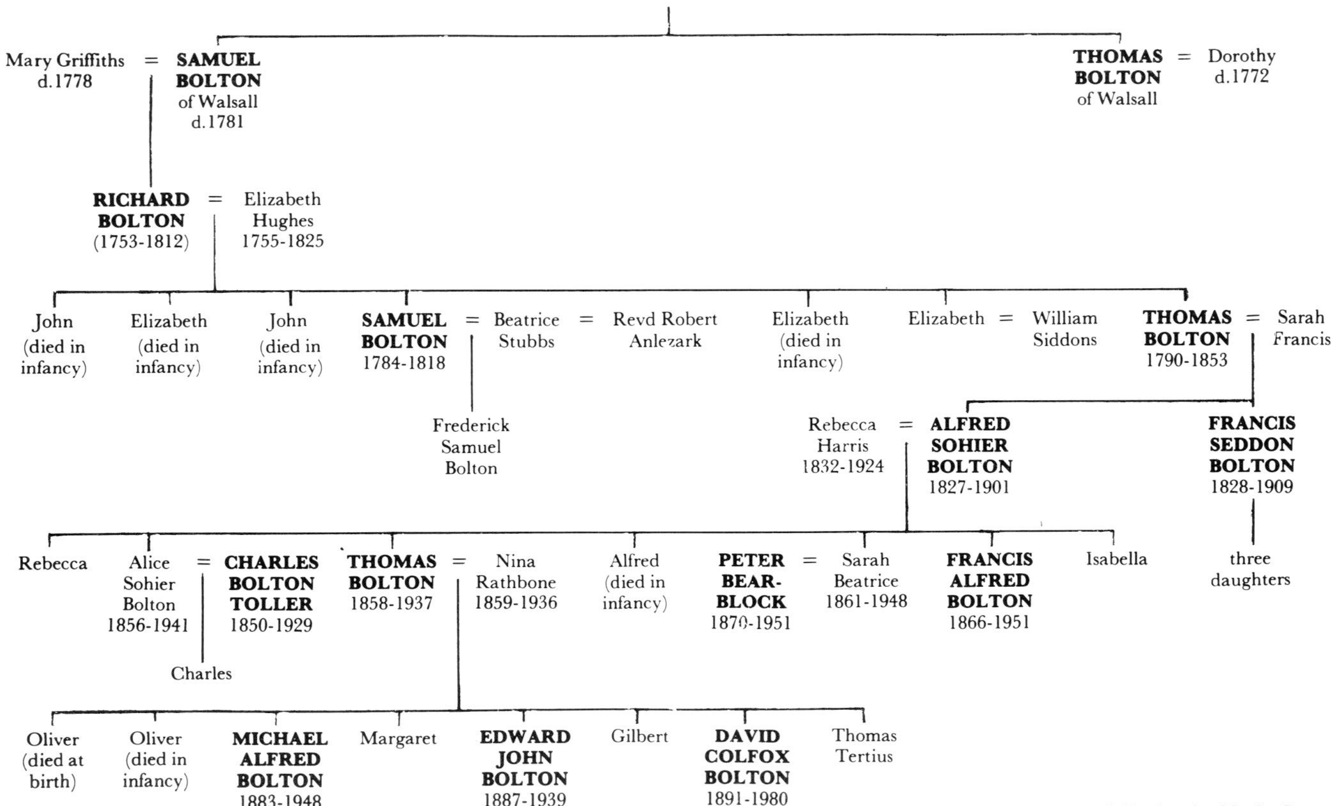

BOLTON = involved in the firm

with 30,000 inhabitants. By 1785 this had grown to 52,250. Birmingham was built on brass, absorbing 1,000 tons in 1780, and 20,000 by 1865. The trade had been given a great impetus around 1770 by the introduction of a new method of making the innumerable metal objects in which Birmingham specialised. Up to that time these had been made largely by casting the metal in small moulds. Now they could be made by stamping shapes out of sheet metal with steel dies. Once the die had been made, a delicate and artistic pattern or shape could be repeated endlessly, and of course cheaply. This led to a great increase in the demand for brass and copper sheets, and the building of a new brass works in 1781, in what is still called Brass House Passage, to provide a local source for the town's workmen. This was made necessary by the copper smelters' action of forcing up their prices as much as £20 a ton. It was the business of fabricating copper and brass sheet which Thomas Bolton decided to take up when he later established the metal works in Broad Street.

It was an earlier Thomas, the brother of Samuel, who seems to have been the first to bring the family to Birmingham. He is mentioned in December 1782 as being in partnership with William Clark as a plater and buckle-maker. This combined a Walsall trade with a Birmingham product, silver plate. Originating in Sheffield, and often known as 'Sheffield Plate', Birmingham had made the manufacture of silver plate its own. A thin plate of silver was bonded onto a sheet of copper by placing them together in a furnace with a borax flux in between, and then rolling the two plates together. Although called a plater, as were others, Thomas Bolton was almost certainly concerned only with the making of buckles from sheets of silver plate. This he would have done by stamping out the shape using carefully prepared dies. The chape or fastener was made separately, and may have been bought in. In Walsall, chape-making had become a specialised trade of its own. Buckle-making was essentially a workshop process, of course, rather than an industrial one using large sources of power.

Thomas Bolton's partnership with William Clarke had been dissolved in 1784, when Thomas was living and working at 26 Edmund Street, in Birmingham's town centre. Thus the strong family tradition that the firm was established in 1783 seems to be borne out, although the nature of the family's activities changed after that date. The late F.A. Bolton declared in 1949 that he had been told '60 or 70 years ago that the business was started in 1783'. An article in the *Railway Supplies Journal*, written after a visit to the company's works in 1896, also states that the firm was established in 1783, clearly on the basis of information supplied by a member of the family.

Thomas continued in business till August 1785, when he went bankrupt, and his stock was sold on 10 October. But his place was soon taken by his nephew Richard, now 35 years old. He had married Elizabeth Hughes in 1779 at Dudley, and appears in Paradise Street in 1787, continuing to trade there as a plater till at least 1791. By 1792 he had formed a partnership with John May, but the partnership was dissolved in the October of that year, and the business continued as 'Richard Bolton and Co'.

By 1794, Richard had moved to larger premises at 52 New Street, a few doors away from the New Theatre. This obviously marked a step up for his business, as New Street was one of the main thoroughfares of the town, and contained many of the shops and warehouses of the more celebrated manufacturers. He had entered into another partnership, this time with James Luckock, the brother of Joseph Luckock, another Birmingham 'toy' manufacturer. James had first joined his brother in his trade as buckle-maker, after being apprenticed as a plater.

Many of the Birmingham 'toy trades', as they were known, were vulnerable to changes in fashion, and buckle-making was no exception. In the 1790s, the trade declined rapidly, as the 'effeminate' shoe lace replaced the shoe buckle. A petition to the Prince of Wales in 1791 by the buckle-makers led to a ban on the use of shoe laces in his household, but in this case the tide of fashion was not to be turned, and by the end of the century, a staple trade of the town had died out. Richard's partner James Luckock became a manufacturing jeweller. Richard himself turned from manufacturing and set up as a merchant.

Richard Bolton was still referred to as a plater up till 1801, and

In Memory of

RICHARD BOLTON

who died April 22nd 1812

Aged 59 years

SAMUEL BOLTON

died Decr the 12th 1818

Aged 34 years

ELIZABETH BOLTON

Wife of the above *Richd Bolton*

who died November 27th 1825

Aged 70 years

The Bolton's gravestone in the Old Meeting House Burial Ground, Birmingham. (From Catherine Beale, Memorials of the Old Meeting House and Burial Ground, *1882.)*

continued to trade from the address in New Street. But in 1803, he became 'Richard Bolton, merchant' and began a new career as a trader and dealer in Birmingham's vast variety of metal products.

Richard continued in New Street until 1806 when he moved his business to Newhall Street, as 'Richard Bolton & Co., Merchants'. In 1810 the business was moved back to nos 83-85 New Street. Richard's home was on Summerhill, at the edge of the town on the road to Dudley and then at Hamstead, a small village to the north-west, not far from James Watt's house, 'Heathfield'. This suggests that he was becoming a man of some substance, and this will be borne out shortly.

Like many others in Birmingham, Richard belonged to the Dissenting community. Their religious activities were centred on the Old Meeting House in Bull Street, which was the parent church of non-conformity in the town. By the eighteenth century it had become predominantly Unitarian. Many of the leading intellectual figures in the town were connected with the chapel, including Joseph Priestley the discoverer of oxygen. Many of its members held radical political views, and supported the French Revolution of 1789. This lead to the notorious 'Church and King' riots in 1791, in which a mob from the town, inflamed by drink and patriotism burned down the Meeting House and also Priestley's home. Thomas Bolton was a seat-holder in the chapel in 1791, as was Joseph Luckock. The Meeting House was rebuilt and re-opened in 1795. When he died in 1812, Richard Bolton was laid to rest in the burial ground nearby. His gravestone survived till 1881, when the property was sold to the London and North Western Railway to make room for an entrance-tunnel to Snow Hill station.

Richard Bolton was only 59 at his death, but he had come a long way from his humble inheritance of 1781. The measure of his success can be found in his will. He left a wife, Elizabeth, who was to survive him by thirteen years, an elder son Samuel, aged 28, a younger son Thomas, 21, and between them in age a daughter Elizabeth.

From the sale of his business effects, Samuel and Thomas were to receive £1,500 each in cash, another £1,000 was to be invested for daughter Elizabeth, and £3,000 to provide an income for his widow, a total of £7,000. These effects included all his business assets, but not the premises which were leased from the Ryland family. The will also shows that Richard already had connections with the copper trade, for he left his four shares in the Crown Copper Company to his wife. The Crown Copper and Spelter Company was set up by Birmingham interests in the late 1790s to smelt copper, and built a works at Neath.

The merchanting business had clearly done very well in the few years since Richard had given up the dying trade of buckle-maker. This was due almost certainly to the war-time conditions which prevailed before 1812. Demand for copper was high, particularly for the Navy, and shipyards were prepared to pay a high price for it. The merchant was in a favourable position to benefit from inflated prices during the wars against Napoleon.

Thomas Bolton, 1790-1853.

After Richard's death there was a brief interregnum. Although Samuel was 28, he was not given control of the business, and indeed he died shortly afterwards in 1818, being buried next to his father in the Old Meeting House cemetery. After Richard's death the business was carried on by a Joshua Bolton, about whom nothing is known at all. He may possibly have come from Richard's uncle Thomas's side of the family. He could have been asked to take charge, because Samuel was unable to for some reason, possibly ill health, and Thomas Bolton being still only 21, was thought to be too young and inexperienced.

Thomas's chance was soon to come however, because of his brother's early death. By 1818, he was already listed in the Directory with his brother for the first time. In 1820, now aged 30, he was in full charge of the premises at 83-85 New Street, and all the commercial transactions, including the endorsing of bills. The era of 'Thomas Bolton & Co' had begun.

Under his management, the business soon expanded, and was moved to more spacious premises at 109 New Street in 1823. The site has been redeveloped several times since, so that no trace of the original buildings remains. Thomas Bolton continued to trade from this shop and warehouse till 1845.

The following year, 1824, Thomas married Sarah Francis of Edgbaston, and settled down in this newly-developed suburb of Birmingham. They lived at 10 Calthorpe Street until a few years before Thomas Bolton died. Again there is no remaining building of the period, as the site is now occupied by a modern banking hall. The area was favoured by many of the well-to-do citizens of the town, since only residential development was allowed on the Calthorpe Estate.

In 1825, his mother Elizabeth died, at the age of 70, leaving the residue of her estate to Thomas as the sole male survivor, apart from her seven-year-old grandson, Frederick Samuel Bolton, the son of Samuel and his wife Beatrice. Frederick became a clergyman in later life, and while vicar

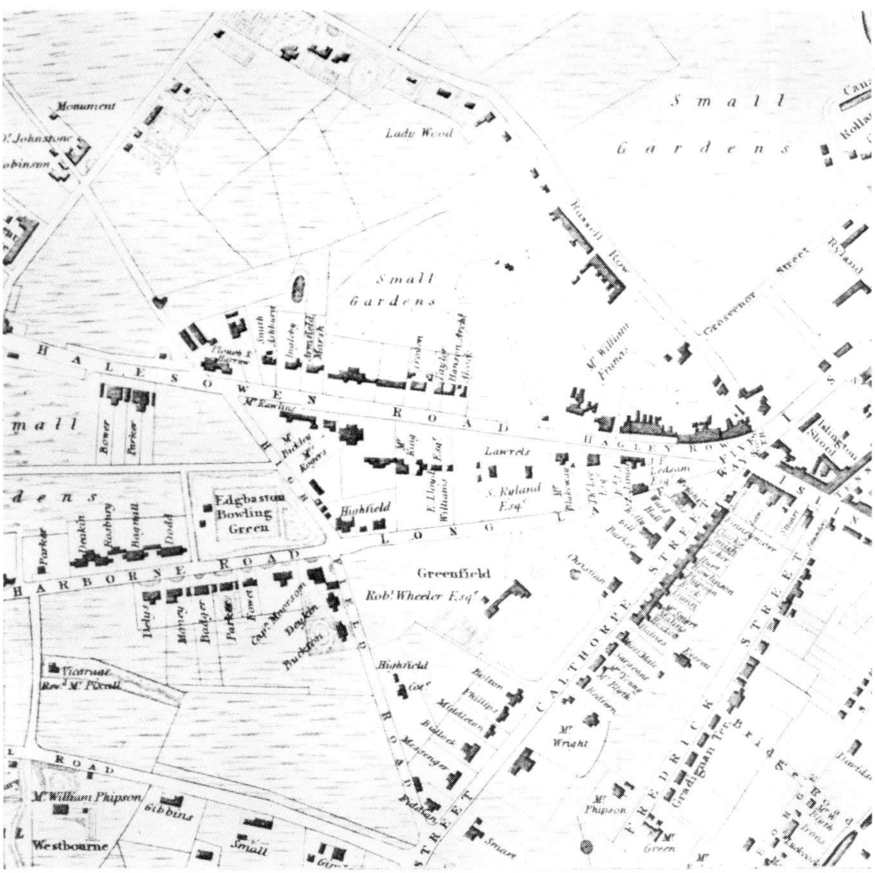

Plan of part of Edgbaston, Birmingham, in 1834 showing Thomas Bolton's home on Calthorpe Street (now Calthorpe Road). This plan is in fact incorrect as Bolton lived at number 10, which was on the opposite side of the road and nearer to Five Ways.

of Salt in Staffordshire, conducted some researches into family history, from which much of the family tree before Thomas Bolton is derived. Elizabeth inherited the family Bible from her mother, in which she had written, as was the custom, the details of the births of her children. An invaluable guide to the genealogist, the fly-leaf of the Bible also recorded the private tragedy of Thomas Bolton's parents: the first three of their children died in infancy, and another baby Elizabeth had died before the surviving daughter was born.

In the 1820s 'Thomas Bolton & Co' began to acquire a reputation as traders with the USA, an important and rapidly growing market. The firm was named as a leading house for Birmingham firms wishing to sell there, in a contract between a Birmingham lamp manufacturer and neighbour of Thomas Bolton in Calthorpe Street, Thomas Messenger, and his Boston agent. The American connection became a personal one, as well as a matter of business. Thomas's legal and commercial representative in Boston, the principal port for the North-Eastern states and the hinterland, was William D. Sohier, a lawyer of considerable standing. He became the godfather of Thomas and Sarah's first son, Alfred, born on the 20 February 1827, who was also given the middle name of Sohier. William D. Sohier presented a cup to the young Alfred in 1834, inscribing it as having been given 'by his father's honoured and esteemed friend'. The cup is still in the possession of Alfred's grandson, Mr Martin Bolton. The Boltons seemed to have made good connections in the United States throughout the nineteenth century, to their own advantage. This resulted in a visit to many East Coast and Mid-Western non-ferrous metal works in 1889, where many machines of the Boltons' devising were in use.

Thomas now seemed well set for a profitable career as a national and international trader in Birmingham metal wares. Many of the town's traditional markets were still active. Despite the ending of the slave trade, or perhaps because of it, the West African market was absorbing large amounts of the traditional 'Guinea Rods', thick brass wire rods about three feet long. These were no longer exchanged for people, but for tropical products increasingly in demand, particularly palm oil. Considerable quantities of brass sheet went from Birmingham to India, where brass cups and other items were used extensively in Moslem ceremonies.

But in the 1830s new developments were emerging in Birmingham's effervescent economic life, which gave impetus to the growth of copper and brass fabrication on an even larger scale. The first was the introduction in 1832 of 'Muntz metal', otherwise known as 'yellow metal'. This was a 60-40 per cent mixture of respectively best selected copper and foreign-zinc. It entirely superseded the copper sheathing used to cover the bottoms of merchant ships, and made a fortune for its patentee, George Frederick Muntz, which he admitted to have been at least £68,000. It was considerably cheaper than copper sheathing and it was also claimed that it kept the bottoms of ships cleaner. This brought a great demand for sheets of the new material which were made by rolling at red heat from cast ingots. The ingots were produced by the patentees, but the rolling of plates was done by other firms. By 1860, Birmingham was making 11,000 tons of yellow-metal plates a year.

The second development came as a result of the new phenomenon of the age, the railway and its steam locomotives. The age of railway construction which began in earnest after 1830 when the success of the Liverpool and Manchester Railway created enormous new demands for copper and particularly brass tubes. George Stephenson's *Rocket* of 1829 which introduced the tubular boiler in its modern form, had twenty-five copper tubes, each 3in wide. The boilers of later locomotives came to have as many as 150 tubes passing through them, and by 1850 there were probably 3,000 locomotives running in the United Kingdom. Each tube could weigh up to 25lb, and would be replaced at least every three years. Up to 1838, only soldered brass or copper tubes could be used. These were originally made by cutting a strip from a sheet of metal and hammering it round an iron bar with a diameter of the size required for the inside of the tube. The joint was then soldered. Draw benches for tube-making were introduced into Birmingham at the beginning of the nineteenth century, and the production of tubes put on an industrial basis. Brass strips, partly curved and with a flange at the end, were drawn through a die to produce a cylinder with a seam running down its length. This was then soldered in a stove which heated the tube to a red heat. The soldered tube could then be drawn down again if necessary using a mandrel inside to give the correct diameter. These tubes were excellent for many purposes, and Birmingham produced them in large quantities. But the steam locomotive boiler, operating at high pressures and temperatures created a need for solid, seamless tubes, which would not burst. Exploding boilers were one of the most spectacular and undesirable hazards of early railway travel. Railway companies would go to great lengths to avoid them. The first successful solid, or seamless, brass tubes were made in Birmingham in June 1838 by Charles Green. He made them from short, thick cylinders of brass, which were drawn down by a special tool, helped by repeated annealing. By 1860, Timmins states that 6,500 tons a year of seamless tubes were being made in Birmingham, of which 4,500 were for locomotives, and the rest for marine engine boilers.

These two developments did much to change the industrial scene in Birmingham, and must have weighed heavily in the thinking of Thomas Bolton in the 1830s. The outcome of this was a decision to start producing goods as well as selling them. In September 1838 in partnership with Thomas Clowes, he acquired the lease of the 'mill and buildings' at 13 Broad Street, which became known as the Birmingham Metal Works. At a later date, their neighbours at no 14, on the corner of Broad Street and Baskerville Place, were Nettlefolds the screw manufacturers. The products of the Birmingham Metal Works were rolled sheet copper and brass, tubes and wire. Within a year, Thomas had installed a Boulton and Watt-type condensing beam engine, using low pressure steam. This was

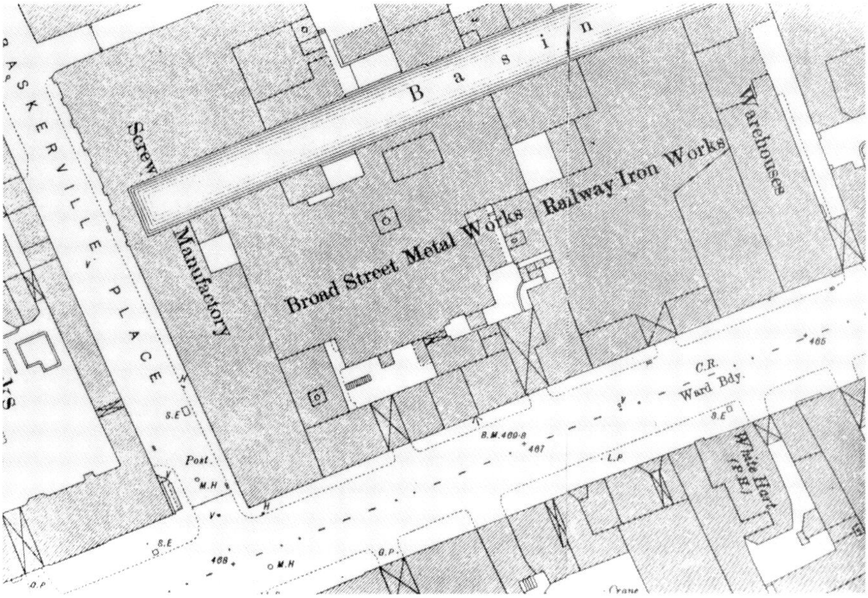

A plan of Thomas Bolton's Broad Street Metal Works, Birmingham, about 1903.

The frontage to the Broad Street Metal Works, Birmingham.

the source of power for the rolling mill, and drawing benches. The firm was known as 'Thomas Bolton & Co'. By 1841, Thomas Clowes had dropped out, and the premises had been further enlarged, to include a wharf, a house for the manager and other additional workshops.

On the trading side, Thomas had taken on a new partner in 1841, Samuel Buckley. His own interests were moving more and more towards the industrial aspect of his activities, and in 1845, he referred to himself in the *Post Office Directory* as 'manufacturer of rolled metals, wire, tubing etc.' In the same year he sold out his interest in the merchant house he had built up over 25 years, leaving Samuel Buckley to continue on his own. This was a major break with the past, and with a business which had established him in the town. By now he had been an alderman for four years, and a magistrate since 1839. He was one of the signatories of the proclamation of intent to prosecute those who organised mass meetings in Birmingham at the time of the Chartist riots in 1839, as being 'contrary to law and dangerous to the tranquillity of the borough'. He was now living in the fashionable suburb of Edgbaston with his wife and mother-in-law, and his two sons, Alfred and Francis Seddon. Both boys went to the Proprietary School in Edgbaston, and then to University College, London. The dissenting community in Birmingham had strong connections with the college, which paid particular attention to the sciences which were neglected by Oxford and Cambridge. However, Alfred had to leave his studies in London before taking his degree, and return to Birmingham, because of the ill-health of his father. As the elder son, it was to be his responsibility to take over the leading position which Thomas Bolton occupied up to that point.

On his return to Broad Street, Alfred was to find the copper trade at a fundamental turning-point in its history. It was in the 1840s that electricity began to make its appearance on the industrial scene. At this early stage, the impact came firstly from the development of the dynamo, and secondly from the introduction of the electric telegraph. The dynamo owed its origins directly to Faraday's work. His experiments at the Royal Institution in London between 1821 and 1831 had shown the possibility of converting mechanical motion into electric current, and vice-versa, through electro-magnetic induction. Birmingham had seen the first significant industrial application of this new source of continuous electric current in 1842. The Birmingham firm of Elkingtons had been experimenting for some years in electroplating metals with silver, using current generated by a 'Voltaic pile' or battery. J.S. Wolrich of Birmingham patented a dynamo in August 1842, or 'magneto-electric machine' as it was then known, which Elkingtons and other firms used as their source of electricity, for depositing silver onto cheaper metals. This innovation, in truth, was probably less decisive in the triumph of electroplate over the traditional plating methods, than the overcoming of the chemical problems associated with electro-depositing. But it was a notable new feature in Birmingham industry in the 1840s. The Wolrich machine was of course quite small, and the significant development of electro-deposition for the mass-production of refined copper itself depended on further advances in dynamo design, which lay some years ahead.

The most immediate impact on the young Alfred was made by the development of the electric telegraph. The continuous electric current produced by electro-chemical methods in the 'Voltaic pile' had for some time been applied to communications. The possibility of sending signals down a wire had been realised early in the nineteenth century. Interest in the telegraph had been stimulated further by the growth of the railway, and the first full-scale experiment in Britain was carried out in 1837 on a line laid between Euston and Camden Town stations, stations belonging to London and North Western Railway Company. This used copper wire and was highly successful technically, but for some reason the company took no interest in it and asked for the line to be removed. It took a murder and a man-hunt to bring the electric telegraph back into the public eye. Two years after the Euston-Camden experiment, the Great Western Railway had agreed to install a telegraph between Paddington and West Drayton, a distance of about thirteen miles. It was this that led to the arrest of John Tawell in 1842. He had murdered his lady friend Sarah Hart at her home near Slough and taken the slow train from the Great Western Railway station to London. When the police discovered that a man answering his description had boarded the train a message was sent to London by the telegraph. As a result, Tawell was arrested, found guilty and hanged. It seems that the Victorian fascination with sensational murders, particularly one involving such an unusual method of capture, did more for the telegraph than the scientific marvel of almost instantaneous communication. The Electric Telegraph Company was set up to exploit the new wonder, and by 1850 they had erected 2,215 miles of telegraph line in Britain.

Some of these early telegraphs used copper wire. But most of the new overland telegraphs particularly those carried on poles were made with iron wire. This was stronger, cheaper, and adequate for conducting the simple signals used with the help of the relay or repeater developed by Charles Wheatstone in the 1840s. A vast order for iron telegraph wire was placed by the East India Company in 1853, said to have amounted to 600,000 miles of wire. Iron wire, galvanised to prevent corrosion was used universally in the 12,000 miles of the United States telegraph system by 1850. There the last of the earlier copper wire lines had been taken down in 1846. In France the entire copper telegraph wire along the roadside from Paris to Rouen was removed in 1848, although in this case by thieves: the relatively high value of copper wire at that time, about £110 a ton, made it worth the risk of such a highway robbery.

The booming market for overhead telegraph wire was thus lost to the iron wire-makers. But copper wire really came into its own with the growth of the submarine telegraph. Here good electrical conductivity had priority, as no relays were possible. Mechanical strength was not important, and a submarine cable was safely out of the way of robbers, or nearly so. For the submarine telegraph there was no alternative to copper. The railway companies already favoured the use of copper in tunnels in order to combat the dampness and acidic secretions from the earth which would quickly destroy an iron wire. One such copper wire, covered with the newly-discovered gum, gutta percha, and intended for a railway tunnel, was used to make the first experiments in underwater telegraphy which were carried out in Dover Harbour in 1849. The success of these experiments was to lead to the Boltons involvement with the very first submarine telegraph which was laid under the English Channel.

In 1850, the English Channel Submarine Telegraph Company was formed with a capital of £2,000, subscribed in shares of £500 by J.W. Brett, the leader of the enterprise, Charles Fox, Francis Edwards and Charlton J. Wollaston. The last of these, the nephew of W.H. Wollaston who had studied the chemical effects of static and current electricity, was made electrical engineer and was sent off to order the copper wire. By this time Alfred Bolton had returned from his studies in London and was present when Wollaston appeared at the Broad Street works. Alfred later recalled the encounter in order to show how inadequate was the technique of wire-drawing at that time to the tasks now being presented to it. Wollaston was imperative that the wire, of 14 Birmingham Wire Gauge, or 0.083in diameter, should be made in continuous lengths, each weighing 30lb. That would have meant unbroken lengths of about 500yd. The foreman was sent for to receive instructions for the order, and when he heard what was required he said 'Does the man think I am a fool?', so impossible did he consider such a job to be. At that time, Alfred Bolton commented, the only wire produced of that gauge came in bundles weighing 4lb, equivalent to about 80 yd of 14 gauge wire.

The order, however, was put into effect, and in this instance it was the view of the foreman which prevailed. The wire was delivered to the Gutta Percha Company in 100yd lengths, to be covered with gum to a diameter of ½in. The lengths were joined by the simple method known as a 'bell-hangers twist', that is by twisting the ends of the wire round each other and then covering the splice with soft solder. No notice was taken of the conductive qualities of the copper as none were specified. The joins were not tested separately for electrical efficiency. Only after a suitable length of cable had been made, was a crude test carried out by immersing it in water, at no specified temperature, and measuring the current produced by a 24-cell battery with a vertical galvanometer. If no serious deflection occurred, it was passed.

The cable was to run from Dover to Cap Gris-Nez. A horse box in the railway yard at Dover served as one shore station and the lighthouse on Cap Gris-Nez as the other. The two shore ends were already in place when the laying of the 25 miles of cable began on 23 August 1850. By 6pm the buoy holding the French end of the cable had been reached. The end of the cable was taken into the cabin of the ship and tested before being spliced to the shore end. Some communication was had with Dover and it was thought safe to connect up the cable and drop it overboard. There was a rush to the lighthouse, but disappointment followed. After many hours of wrestling with the equipment, not a trace of a signal could be heard. The line was dead. The cause was thought to have been damage to the cable from the rocks off the French coast. Another possible explanation emerged in the following year when the *Illustrated London News* reported that a Boulogne fisherman had caught a strange snake or piece of seaweed in his nets, which seemed to have a thread of gold running through it. He cut a piece off and brought it home with him.

Although a failure, the experimental cable whetted the appetite for success. The lessons of the failure were considered, and it was agreed that there should be four copper wires next time, each of 0.065in diameter. These were to be separately coated with gutta percha, and then individually wrapped in tar-saturated yarn. The whole 'core' was to be similarly wrapped, and then protected with ten 0.25in galvanised iron wires. The manufacture was delayed by litigation over the patent rights, and was not finished till 17 September 1851.

There is no direct evidence that the copper wire for this cable was made at Broad Street. Alfred Bolton's diary for 1851 makes no mention of it, although it was first opened for public messages on 13 November 1851, and firmly established submarine telegraphy. It is possible that the wire was made at Oakamoor, as there are several references to a Channel cable being made there. But whether or not the Boltons received the order, there is no doubt about their interest in developing this side of the business. Alfred Bolton was said to have been 'one of those who early discerned the great part electricity was to play in our modern life', and this was soon apparent at Broad Street.

As yet, the electrical qualities of copper wire for the telegraph were not considered. The commercial importance of conductivity was to be the matter of fierce debate in the 1850s in which even scientific giants like William Thomson and Michael Faraday found themselves on opposite sides. The Boltons' concern was to be able to produce improvements in the form of the wire — that is mainly in making longer lengths. They had not been able to meet the requirements laid down by the Channel Submarine Telegraph Company in 1850 and this was clearly a challenge, particularly to Alfred.

By August 1851, he was at work expanding the wire-making facilities at Broad Street. The wire casting shop was extended and in September new fine wire drawing-blocks were put in. Along with improved machinery went a demand for more power for the wire-drawing benches and the rolling mill. It was decided to install another steam engine, this time one of the high-pressure type which were then supplanting the older low pressure condensing engines first developed by James Watt. Alfred did not dispense with the existing engine which was of the condensing type. Always keen to try something new, he set both engines to work in tandem, with the exhaust steam from the new high pressure engine being used by the low pressure one. He also considered the possibility of working them together driving the same shaft. The new engine was erected in November, and its beam put in place on 7 December.

The Boltons now had one of the most up-to-date wire-drawing shops in Birmingham. W.C. Aitken described graphically the kind of improvements that were made in this trade in the Birmingham area since the beginning of the century:

> For the creechy rolls, the primitive draw-benches, and clattering wire blocks, rolls of great size and power, fitted into massive iron frames with adjusting indexes, draw-benches with admirably fitting trains of wheels and chains of great strength, and wire-drawing blocks well-fitted, revolving truly and silently, have taken the place of their imperfectly constructed representatives in a preceding generation.

It was by these means that the first steps were taken by the Boltons to meet the demands of the new telegraph system. As yet there were no changes in the basic methods of production. Copper sheet was still rolled, slit into square rods and these were still drawn down to round rods in the way already described. The innovation of rolling the square rods until they were suitable for the wire-drawing benches was still to come in the making of copper and brass wire.

Amid all this activity, the advertisement of the sale by auction of the entire Cheadle and Oakamoor works of the Cheadle Copper and Brass Company attracted little attention. When no acceptable offer was made at the auction on the 26 September, the owners decided to canvass for buyers. The Boltons were approached in the New Year by their representative Taylor, and Thomas Bolton's interest was aroused. Although the brass and copper works at Cheadle were also for sale, it was the Oakamoor works and more particularly its wire-drawing facilities which attracted him. He and his son Alfred went to visit them in February 1852, accompanied on the 18 February by Francis Edwards and Charles Wollaston, both directors of the Channel Submarine Telegraph Company. It is clear from this and from the way that Oakamoor was run once the Boltons had taken over, that their main aim in buying the works was the expansion of the manufacture of copper telegraph wire. The account books show clearly that this was the main product, and the brass wire which has been the concern of the more traditional previous owners, was very much pushed back into second place. The Boltons needed these new facilities because the scope for expansion at Broad Street was slight. The site was hemmed in on both sides by other industrial premises, whereas the workshops at Oakamoor were by the standards of the time large and spacious.

The Oakamoor works were again advertised for sale in the *Mining Journal* of 14 February 1852. The auction was fixed for 2-4 March. But Alfred Bolton met the Cheadle Company's lawyers on 23 February and made them an offer of £7,750 for the Oakamoor works, its stock and machinery, and the cottages and other property nearby. The offer was promptly accepted the next day, and the agreement drawn up and signed on 1 March 1852, in the names of Thomas Bolton and J.W. Patten. The freehold was bought by Thomas on 1 May for a further £5,000.

The Boltons' move into industry was now signed and sealed. But Thomas himself was in failing health. For the last two years of his life he suffered from a disease of the bowels which brought on his relatively early death at the age of 64. He made out his will in July 1852, and in the following June had a partnership agreement drawn up between himself and his two sons. In this he spoke of himself as having traded for some time 'as a metal dealer and manufacturer of rolled metals, wire and tubing' in a leasehold premises in Broad Street, Birmingham, and freehold mills at Oakamoor. His capital on 31 December 1852 is stated to have been £10,500 but the capital of the two sons is not revealed. He was to be free to attend to the company's business or not as he wished, but Alfred and Francis were obliged to devote all their time and energy to it. Nor were they free to withdraw their capital from the partnership. Each was to receive 5 per cent return each year on his investment, plus any residue. By this agreement, on 30 June 1853, 'Thomas Bolton & Sons' came into being.

Thomas did not long survive the signing of the agreement. In the autumn he became ill, and in November much worse. The end came at 12.30pm on the 5 December 1853. On the day of his funeral, 10 December, the Birmingham and Oakamoor works were closed. Public tributes to him mentioned his 'sincerity and singleness of character', his 'kind and courteous manners', and stressed his business integrity, which 'upheld . . . abroad as well as in his native country, the high character of the English merchant and tradesman'.

Sarah Bolton, Thomas's widow, was left £3,000 and the contents of their house at 24a Calthorpe Street, plus an annuity to be met from the income of the two works. Thomas's nephew Frederick Samuel Bolton received the sum of £500. Everything else, including the business and the property belonging to it, was left to Alfred and Francis to enjoy as 'tenants-in-common' but its value was not specified in the will.

Thus the responsibility for the firm very quickly fell on the youthful shoulders of Alfred, aged 26, and Francis, aged 25. Alfred took charge of Oakamoor, as he had already begun to do before his father's death. Francis was to stay in Birmingham for the rest of his long life, and run the Broad Street end of the business. For a while, this was still the major concern, but very quickly Oakamoor came to be the dominant end of the partnership, and Alfred the leading partner. The main contribution made by Thomas Bolton and Sons in the next fifty years to copper-making, and the electrical industry, came through the initiative of Alfred and his sons.

CHAPTER THREE

Victorian Enterprise 1852-80

Thomas Bolton's exhibit in the Great Exhibition of 1851 was described in the catalogue as follows:

> No. 353. Thomas Bolton. Broad St. Metal Works, Birmingham. Sheet brass, German Silver and copper; specimens of the process of manufacture of brass wire, round and shaped brass and copper wires, and of tubing. Brass and copper tubes, locomotive and mandrel drawn tubing.

There were many strings to the company's bow as a general fabricator of non-ferrous metals. But there were also many other general fabricators in the same field, particularly in Birmingham. What was to distinguish Thomas Bolton & Sons from the rest in the years that lay ahead was the decision to specialise in the electrical applications of copper. In the thirty years after the exhibition and the acquisition of Oakamoor that meant primarily copper wire for telegraphs, and above all for the submarine telegraph.

This market did not amount to much in purely quantitative terms. In the first decade, after 1851, just over 600 tons of copper went into the submarine cables which were laid. But its qualitative significance was enormous. The effect of the telegraph on the development of the world economy and life in general goes without saying. But it was equally important for the development of the electrical industry. It concentrated some of the finest scientific minds of the age on the vital question of the improvement of the electrical qualities of copper. This was of course mainly a problem for the copper smelters, but by specialising in the field of submarine telegraphs, the Boltons became closely involved with the problems of selecting the right materials for the job. They also had to overcome the technical problem of making the wire in the longest possible lengths, to improve both the mechanical and the electrical qualities of the cable conductor. They were limited to some extent by commercial considerations in this effort. Since the telegraph work, although important, was only a part of their output, and a relatively small one, it did not justify the expense on new capital equipment which was needed to overcome the problem. This is probably why the copper and brass wire manufacturers lagged behind the steel and iron wire makers in the introduction of new methods of wire making, and particularly in the rolling of rods directly from wire bars. By 1880 this was the general practice in iron and steel wire mills, but the Boltons were still rolling the wire bars into coil, slitting this into square rods and only then rolling these through grooved rolls into round rod ready for wire drawing. Alfred Bolton noted in 1883 that these methods were different from those used in making iron wire. He was later to pioneer the rolling of copper rod direct from the copper bar or billet, but not until the market for long lengths of copper had greatly increased after 1880, and thus made it worthwhile to invest in new methods. Large grooved rolls had incidentally been used even in Patten's time but only to take the corners off thick square rods. They were known as 'rumple rolls'.

Land telegraphs in Europe and North America were largely made and erected by companies in the country concerned. But the manufacture and laying of submarine cables was a British pioneering enterprise and was for many years entirely in the hands of a handful of companies of which Thomas Bolton and Sons was one. According to William Preece, the Post Office's electrical engineer and close associate of Alfred Bolton, the firm were responsible for all the copper in the early cables of the 1850s. Between 1851 and 1861, thirty-one shallow water cables were laid totalling 3,074 nautical miles, and twenty-one deep water cables, with a combined length of 8,290 nautical miles. These included cables for the French, Dutch, Danish and Swedish Governments. Even as late as 1868, the Boltons were making wire for the French Atlantic cable.

The extent of the wire-making operations at Oakamoor in the first few years can be seen in the following table, drawn from the accounts of the rolling and slitting mill and the wire drawers between 1852 and 1858. The amount of copper used in the conductors of all the cables laid in each year is shown for purposes of comparison with the output from Oakamoor.

Year	Copper used in submarine telegraphs laid (tons)	COPPER			BRASS	
		Finished wire (tons)	Rods made at Oakamoor (tons)	Fine wire (tons)	Wire (tons)	Rods (tons)
1852	3	55	0	.3	17	20
1853	24	239	16	3.7	118.5	144
1854	21.3	147	30	5	120	126
1855	9	61	45.5	7.4	103	120
1856	4	68	76.5	9.1	113	126
1857	121	129	149	67	91	119
1858	83	51	?	?	37	?
Total	265.3	750	317	94.5	599.5	655

Note: The figures for 1852 are for Aug-Dec only.
The figures for 1858 are for Jan-June only.
? indicates that no figures are available.

The first thing that is clear from the table is the relative importance of copper wire at Oakmoor: a total of 844.5 tons against just under 600 tons for brass wire. The amount of brass wire began to fall steadily after the first few years production. The large amounts of copper wire being made in 1853 and 1854 represent sales for land telegraph wire in the early period before iron wire took over. This declined very rapidly as the switch to iron took place. Some copper continued to be used for underground cables, and in railway tunnels, however. The proportion going to submarine cables grows as the decade passes and grew even more in the years which followed.

Another interesting fact to emerge from the figures is the discrepancy between the amount of copper rod coming out of the slitting mill, and the amount of copper wire actually being made. Not until 1856 did Oakamoor produce all its own rod. The reason for this out-sourcing is not clear, but it may have been that the old rolling and slitting machinery left over from the Patten era was inadequate for making rods long enough for telegraph wire. The rods would have been brought in from the copper works in South Wales, or from Birmingham. The steady increase in self-sufficiency in copper rods would thus be an indication of an improvement in the performance of the rolling and slitting mill under Alfred Bolton's guidance. The machinery at Oakamoor would have been satisfactory for brass wire where there was no demand for especially long lengths, and all the brass wire was made from rods slit at the works.

The figures also reveal the amount of copper lost in the course of drawing the rods into wire. The difference in weight between the rods and the finished wire averages about 15 per cent. Most of this, about 12 per cent of the original weight of the rods, was due to scale formed during the annealing process. This was collected and treated at the works to extract the copper, or sold to the Keys at Whiston.

Restarting Oakamoor

Immediately the sale of the works had been ratified on 1 March 1852, Alfred Bolton was put in charge. He sometimes went with his brother or Thomas Bolton, but the main responsibility was his. During March he went with Copely Chambers a Birmingham engineer to clean the steam engine which, together with four waterwheels, powered the works. This large beam engine had been installed about 40 years before and continued in use into this century. He also took Fisher, a Birmingham supplier of metal-working machinery with him to find the best positions for wire-drawing benches. By 3 April Alfred was ready to restart some of the former employees in the rolling, slitting and wire drawing departments. Tube-making and other activities had to wait. The key man for wire-making was Richard Wilson, who had been in charge under the Pattens. But he had been the victim of some degree of religious persecution and had left Oakamoor. A strong Methodist, he held

Oakamoor about 1860, showing the works virtually in the state that the Cheadle Copper and Brass Company sold them to Thomas Bolton.

Wesleyan meetings at his house as there was no place of worship in the area for them. Two Anglican ladies in the village had taken great exception to this and brought pressure to bear through his employers so that he was obliged to either give up his religion or his job. Wilson left and went to Holywell in North Wales, where a similar range of industrial activity was carried on as at Oakamoor, and where no doubt Methodism was regarded more favourably. The Boltons however persuaded him to return and continue as head of the wire-making department. Perhaps as a result of this, Alfred, whose own parents were strongly non-conformist, later built a non-denominational chapel in Oakamoor for the free use of the workers. Samuel Haynes was also restored as head of the rolling and slitting shop, and production was restarted. On 2 August the first five tons of copper telegraph rods were noted in the accounts. On 15 September, Alfred was busy sending off wire, and on 24 September a consignment of wire was sent to the Gutta Percha Company to be covered with insulation. All goods went by rail from the siding which the Pattens had built inside the works, but by 24 November 1852 the amount of telegraph wire to be shipped had grown so much that Alfred Bolton had to arrange to have it sent by canal to London. In the first five months of operations at Oakamoor 55 tons of telegraph wire was made and dispatched.

It was often said that the Boltons were able to recover the cost of buying the works by selling the ash and slags from the furnaces. Whether they recovered the full amount cannot be proved now, but in September Alfred Bolton was busy 'getting the ash dirt loaded'. As late as 1863, he was still grinding an old stock of casters' ashes and a heap of refinery slag with a crushing wheel he had brought from Keys after the sale of the Cheadle works in 1856.

1853 saw the largest output of copper telegraph wire at Oakamoor in the early years, a massive 239 tons. There was also more activity in submarine telegraphy: 625 miles of cable were laid, mostly around the coasts of Britain. The first foreign order was made by the Danish Government for a cable across the Denmark Strait. Production was still concentrated on wire. Alfred Bolton was ready to sell many of the tube benches in the old mill. He was still commuting to the works from Birmingham, and began to look for somewhere to live nearby. In October he looked at a house in Cheadle but did not like it, and he did not finally

The strip mill annexe at Oakamoor, containing coal-fired annealing furnaces. The bricks over the doorway were inscribed T B & S 1858. This building was among some of the earliest extensions after Bolton acquired the works.

move until his marriage the following year. Most of his visits were to deal with machinery problems. He wanted stronger wire blocks to cope with the work. The Dover-Ostend cable alone needed 480 miles of 16 gauge copper wire. This was the standard gauge used in submarine cables up to 1856, established by the success of the 1851 cross-channel cable which had used this gauge of wire.

In 1854 another seven submarine cables were laid. Two foreign orders included the largest yet: 660 miles of number 16 gauge for the French Government's line to Corsica. The accounts show that 158 tons of telegraph rods were in stock at one stage. In this year, on 14 June, Alfred Bolton married Rebecca Dickinson Harris, at Edgbaston Parish Church in Birmingham. After the honeymoon, they both returned to Oakamoor to live. They received a rapturous reception from the workforce, then about eighty men, when they arrived at the station. The whole valley turned out to welcome them, with a band playing and arches of flowers over the road and over the gates of their new home, Sunnyside. A large crowd gathered outside the house and the new Mrs Bolton had to show herself at the door many times. This was more than a conventional reception: the Boltons were regarded as the saviours of the works and therefore of the livelihoods of most of the people in the valley. Alfred also involved himself much more with the needs of his employees than the previous owners. He had a school room built in the works in 1856, a reading room in 1857, and started evening classes with a full time teacher in 1859. He himself gave lectures on a variety of subjects in the reading room, often once a week.

In 1855 the tempo of the work slackened, and this gave an opportunity to strip out all the old wire blocks, shafting and other machinery in the old wire mill in preparation for new installations. Major repairs were done to two of the waterwheels. The large steam engine, which had been giving trouble from the start, was finally put in proper working order. Alfred Bolton also began to set up a copper refining furnace. A large drawing-stone was brought from Whiston to the works, as part of the new furnace. He was already aware of the importance of the purity of the copper which was going into the telegraph wire, and that its quality would be decisive for the firm's reputation in this field. W.H. Preece later claimed that Alfred Bolton was encouraged in this search for greater purity by William Thomson, later Lord Kelvin. In 1855 the only activity in submarine cables was the order by the British and Ottoman Governments for cables to assist with the war effort in the Crimea. Some 450 miles of 16 gauge cable with a single conductor were hurriedly made with no armouring to ensure quick delivery. 1855 also saw the opening of Thomas Bolton and Sons first London office at 17 Thaives Inn, Holborn.

Single hole wire-drawing blocks at the top bronze mill, Oakamoor. ▶

Bronze wire-drawing on wire blocks, at Oakamoor. This and the previous illustration show the method of producing wire before the advent of continuous wire-drawing machinery. ▶

The First Atlantic Cable

The breathing space of 1855 and early 1856 was opportune, because it gave time to prepare for the firm's biggest challenge yet, the first Atlantic submarine cable.

By 1856, 1,500 miles of successful cable had been laid underwater. It was considered that the time was now ripe for the most ambitious scheme so far. In 1856 the Atlantic Telegraph Company was formed by the American Cyrus Field, Charles Bright, John Brett, and O.E.W. Whitehouse, a retired medical man who had become interested in electrical matters. They were to lay a cable between Europe and Newfoundland before 1862. They got a guarantee of £14,000 per annum from the British Government to be paid when the line was working, and a similar guarantee from the US Government. £350,000 was raised in £1,000 shares. Finally, and fatally, they undertook to lay the line in 1857. That meant in reality by mid-1857, as the cable could only be laid during the summer months.

During 1856, two shorter cables were made by Glass, Elliot and Co on the Thames, to link Prince Edward Island and New Brunswick, a distance of about twelve miles, and Newfoundland with Cape Breton, which needed a cable 85 miles long. For the first time, a twisted strand cable was used, using seven 22 gauge copper wires. There were earlier cables with more than one conductor, but these had been laid straight in the core, and not twisted. The idea of stranding the conductor was originally William Thomson's, and he had considered patenting it in 1854. The advantage over single conductors was obviously less risk of loosing communication through breakage; a stranded core would also be more flexible, and so easier to coil and handle than a multi-wire core that was not twisted.

Experiments were made with other cables at Glass, Elliot's works, but the form selected finally was the same as that of the Canadian lines, a seven-strand conductor using 22.5 gauge copper wire, weighing 93lb per mile, or 107lb per nautical mile. One advantage of using a thinner gauge of wire was that the wire drawers could produce 22 gauge wire in greater lengths than was possible with the earlier 16 gauge standard. 22 gauge wire had less than a third the sectional area of 16 gauge, and so a given length of rod would produce considerably more wire. Onto the wire strand were to go three coats of gutta percha and the core was to be covered with jute yarn saturated with a composition of tar and other materials. The outer armouring was to consist of eighteen strands of 22 gauge iron wire, with seven wires to each strand. The copper content amounted in all to 119.5 tons.

Charles Bright wanted a much thicker conductor of nearly 400lb a mile, and 'maximum purity' in the copper, but he joined the company after the specifications had been agreed. William Thomson carried out tests on some of the samples of copper from different manufacturers, and found considerable variations in the conductivity of the copper, although the samples from each manufacturer were consistent with each other. He had the samples analysed chemically and found the connexion between the presence of impurities in the copper and the loss of conductivity. Thomson had also been the first to point out that the speed of a signal, and therefore the commercial usefulness of a telegraph line, was dependant on the degree of resistance in the conductor, the famous 'law of squares'. Both men appreciated the importance of conductivity as an essential quality in telegraph wire, and there was much to improve on, as the average for the first Atlantic cable was only about 50 per cent of that of pure copper.

The June deadline for the manufacture of the cable meant that everything was hurried. The cable itself was to be made in two halves, one by Glass, Elliot and the other by Newall's. On 7 January, Bolton went to London to make the arrangements with John Statham, works manager of the Gutta Percha Co, for the Atlantic cable wire to be insulated, and at the same time saw the representatives of copper smelters William Foster and Co, about the copper. He had to produce about 20,000 miles of 22-23 gauge wire in five months, a total weight of 108 tons of copper, and in the longest possible lengths. Alteration to the rod blocks and to the large engine were made in January. In the event, the task proved too much. About 60 tons of wire were made by Boltons and the rest was made by their neighbours in Birmingham, J. Wilkes & Co. The shortage of iron wire proved an even greater problem as over 380,000 miles were needed. By the 12 June the *Engineer* was commenting that 'the daily consumption of fine wire (for the Atlantic cable) has been greater than all the wire-drawers in England could furnish.' The cable was tested for conductivity in lengths of 1 to 1.5 miles. The testing was done under water, but at no fixed temperature. Finally, on 6 August 1857, the cable-laying got under way. The British ship *Agamemnon* and the United States *Niagara* sailed from Valentia Bay on the south west coast of Ireland. 334 miles were laid in three days, but the cable broke as the depth increased to over 2,000 fathoms. This put an end to the expedition for the time being.

Further deep-water trials were made at 1,800 fathoms in the Bay of Biscay. 900 extra miles of cable containing about 37 tons of copper were ordered to replace what had been lost. This time William Thomson forced the Atlantic Telegraph Company to insist on 'high conductivity' copper in the contract.

In 1858, a second attempt was made to lay the cable, using the same two ships, carrying the rest of the original line and the extra cable which had been ordered. This time the ships were to start in mid-Atlantic and then steam in opposite directions. The attempt was successful, and on 17 August the first message was transmitted across the Atlantic. There was naturally great excitement at the achievement, but after several weeks, a fault developed, and the signals became more and more unintelligible. After 700 messages had been transmitted, the line failed completely.

Despite the loss of the cable, important lessons had begun to be learned, particularly in relation to conductivity. After 1858, the copper smelters were brought to the Gutta Percha Co's works to see conductivity tests

Alfred Sohier Bolton (1827-1901), at Christmas 1899.

being carried out on their copper. It was seen that some ores produced more impurities than others, and this began the improvement in the copper brought about through the 'best selected' method of smelting. From that time a branch of copper making grew up called in the trade 'conductivity copper'. By 1861, a cable of 81 per cent conductivity as compared to pure copper had been produced and by the time of the

Rebecca (1832-1924), Alfred's wife.

second Atlantic Cable in 1865, 96 per cent had been reached in commercial copper wire. Alfred Bolton worked with the smelters to ensure that only the best copper was used in his wire. In February 1858, he himself worked in the Oakamoor refinery for a few days, studying the matter at first hand. Another aspect of manufacture over which he could excercise some control was the electrical testing, and in his next major submarine contract this was given a high priority.

Further improvements were made to the wire mill. Alfred began to install a new horizontal steam engine in November 1858, bringing the number of steam engines on the site to three. In March the following year a full-time engineer, James Bourne, was taken on, to be in charge of them.

Finally, on 29 March in this significant year in the firm's history Thomas Bolton II was born, Alfred's third child, and eldest son. He was to follow his father as the leading personality in the company, sharing with him strong interest in scientific matters.

1859 was to be the busiest year yet for the submarine telegraph makers. 310.8 tons of copper wire were made for 2,906 miles of cable. In the first few months of the year, Alfred Bolton was re-equiping the rolling mill at Oakamoor. Rather than buy new machinery, he went in February to Swinford-on-Avon near Bristol, where the copper and brass makers John Freeman and Co were selling off their rolling mill and its machinery. Freeman and Co were incidentally the descendants of the very first copper company which was set up in 1690 at the time of the revival of the copper industry. He sent William Tipper his works manager down a week later to buy some of the rolls and other plant, and set to stripping the old machinery out of the rolling mill. The Swinford machinery arrived on 9 March, and was quickly installed. By 6 April, the shafting of the wire mill had also been altered, ready to be driven by the new steam engine.

Alfred was now ready for the next major contract, the Falmouth-Gibraltar telegraph. The very next day, 7 April 1859, he went to London to see the Gutta Percha Company, the copper company Sims and Co from South Wales, and the cable-makers Silver and Co and on 28 May made his arrangements with the Gutta Percha Company for the Gibraltar telegraph wire. This was made in June under controlled conditions, attempting to overcome in advance all the problems met by the Atlantic Cable by careful attention to design, manufacture and testing. The conductor was thicker than the Atlantic standard of seven 22 gauge wires which had become established for longer cables after 1856. It was to weigh 400lb a mile, over four times as much as the Atlantic cable. Six samples were tested by Dr A. Matthiesen who found them to give 90.7, 89.5, 79, 78.2, 67.4 and 74.4 per cent conductivity. The wire was still being made out of different sorts of copper, but the average of about 81 per cent for the whole cable was a considerable improvement. However, this higher conductivity was bought by quadrupling the weight and therefore the cost of the conductor. Unlike other cables, this one was tested at a constant temperature. The essential connection between temperature and conductivity was to be shown scientifically by the same Dr Matthiesen in the following year. In the event the cable was too late for the 'laying season' in 1859. Eventually it was used in 1861 for the line between Malta and Alexandria.

In 1860, a Board of Trade enquiry began into the reasons for the failure of the Atlantic Telegraph Company's cable. The public interest in the question was considerable, and every leading expert of the day gave evidence. The published report contains over 500 pages. The immediate causes were not hard to find: the haste in manufacture, damage done to the cable in storage and handling in the ramshackle sheds and yards which were used by the cable companies at that time, were quickly identified as contributory causes. But the report went further into the basic principles involved in the long-distance submarine telegraph and its manufacture.

Charles Bright argued that low conductivity was the main problem to be solved. The Atlantic cable had only 40 per cent and upwards, and the core had been too small. The solution lay either in thicker cores or higher conductivity in the copper. He also criticised the weak joints in the wire.

Dr A. Matthiesen presented the results of his investigations into the causes of the different electrical conducting powers of commercial copper. William Thomson had already shown the connection between impurities and loss of conductivity in copper wires. Matthiesen's experiments showed exactly what these impurities were and the extent to which they affected the conductivity of the copper. He found that the most injurious was the oxide formed by the oxygen which copper readily absorbed from the air when it was molten. This could reduce the conducting power by as much as 28 per cent. He recommended that the smelters inject hydrogen into the copper while it was still in a state of fusion to remove the oxides. Matthiesen also found no alloy of copper would conduct better than the pure metal and in all cases he thought that the aim should be the production of the purest possible copper. His experiments on samples of commercial copper gave 72.22 per cent conductivity for bright copper wire, 81.35 per cent for 'best selected' copper, 88.86 per cent for the Australian Burra Burra copper, and 92.57 per cent for the recently discovered Lake Superior copper. His report also pointed out the relation between temperature and conductivity and the uselessness of testing wire for conductivity unless the temperature was constant. His experiments also revealed for the first time that hard drawn copper was 2 to 2.5 per cent less conductive than unannealed copper. This was an important point for Alfred Bolton to bear in mind, as it meant that the mechanical treatment of the wire in his works could reduce the electrical quality of the final product unless care was taken in drawing and annealing.

The report also concluded that twisted strand cable was undesirable, as compared with a solid copper conductor of the same gauge. It was thought that the lengthened path of the current would increase the electrical resistance, and that as the sectional area of a strand was less than that of a solid wire, there was also a loss of conductivity.

Matthiesen's proposal that the smelters should shoot a jet of hydrogen into their molten copper did not bear any fruit in practice. There was no commercially available supply of the gas which would make it practical. So the solution to the problem of conductivity had to be sought elsewhere. However, one lasting effect of his inquiries was the establishment of the 'Matthiesen standard' which expressed the conductivity of a sample of copper as a percentage of that of pure copper, rated at 100 per cent under certain specified conditions. This remained the universally used standard until the improvement of refining and smelting techniques began to overtake it. Before the end of the century, Alfred Bolton was making copper which *exceeded* the 100 per cent standard!

The conclusions of the report gave Alfred food for thought, and he set himself to find a solution to the problems of making an efficient submarine cable conductor. He sought a solution which would combine the mechanical advantages of a stranded wire, with the electrical advantages of a solid conductor. His first solution was a segmented copper wire. This was to consist of four copper wires each forming a segment of a circle, drawn into a copper tube. In 1862 he was busy experimenting with this, with copper supplied by Sims and Co. By 2 December, he had advanced to the point of getting an order from Chatterton, for it to be used on the proposed Persian Gulf line, part of the telegraphic link with India. He went to Birmingham to discuss with his fellow wire-maker Wilkes about sharing a patent, taking samples of the segmented copper bars with him. From there Alfred went to Llanelli, where Sims and Co had their works. For several days he experimented, with the help of Nevill, to find the best way of rolling the segmented copper bars. It became clear that they were too difficult to manufacture, and a new approach had to be tried.

The result of the modifications appeared in the patent number 22, which was filed on 2 January 1863 by Alfred Bolton. His idea now was to have a solid rod of copper as the central conductor, and surround this with one or more tubes of copper or copper alloy. The wire was to be made by rolling and drawing the tubes and the rod down to the required dimensions. In some cases, one or more of the tubes might be segmented into one or more parts, but in all cases the outer cylinder must be entire. Wires made in this way, he observed 'are particularly applicable for electric telegraph purposes'. In January and February the annealing furnace in the large wire-mill was rebuilt to handle this new product. On 19 February Alfred was again in London consulting with Chatterton and Latimer Clark about the segmental wire. Clark was the author of the final report which was mentioned above, summarising the conclusions of the enquiry into the Atlantic telegraph cable. By 12 March Oakamoor was making the patent wire for the Persian Gulf line. The company finally selected a conductor weighing 225lb a nautical mile. The rods were being supplied by Sims and Co, who Alfred saw in London on 24 March. Alfred wanted to make the rods himself at Oakamoor and the next day, he went to look at 'a train of grooved rolls for rolling copper rods'. In April, Nevill of Sims and Co sent his manager James Drew to Oakamoor to discuss the manufacture of rods for patent wire. Alfred went again to London on 12 June to arrange the final specifications of the 'patent telegraph wire', which were passed on 24 June.

In the end the patent wire did not prove successful, commercially. The order was not repeated, and later lines reverted to the stranded wire of the original Atlantic cable. Despite its undoubted electrical merits—the cable averaged 89.14 per cent by the Matthiesen standard—it probably proved too expensive to make. It did however push Alfred Bolton further along the road to rod-rolling of copper wire with grooved rolls.

The Second Atlantic Cable

Almost immediately after finishing the Persian wire in September 1863, Alfred was taking an order in London for 150 tons of wire for a new Atlantic cable. Plenty of time was to be allowed for the manufacture and testing, to prevent a repetition of the previous failure. An 85 per cent conductivity standard was specified, and the wire was required in 38lb lengths, that is about a quarter of a mile of the 18 gauge wire ordered. This was to be stranded into a conductor with seven wires weighing 300lb a mile.

The testing apparatus previously used at Oakamoor was examined by another of Nevill's men from Sims and Co and a new apparatus was sent

Belt driven stranding machines at Oakamoor. Inset: *A twentieth-century example of stranded cable, $2^3/_8$in diameter.*

from the Gutta Percha Company in October. But all was not going smoothly. On 24 November, Alfred was in London seeing the Gutta Percha Company about the Atlantic Telegraph contract which they wanted to alter or annul. He does not say what the problem was, but it may have been to do with the length of wire which was expected. 38lb lengths of 18 gauge copper was considerably more than the 6 or 7lb normally made. Boltons were obliged to make changes in their machinery to cope with this, and the cost of this may have been a source of contention. However, the 'Atlantic matter', as Alfred referred to it in his diary, was finally settled at a meeting with Chatterton in Birmingham on 10 February, 1864.

The total order of some 300 tons of copper wire was again shared with J. Wilkes, with Boltons contracting to make 200 tons of it. The first batch of Atlantic wire was drawn on 3 June 1864, and quantities were drawn regularly every fortnight until 20 June 1865 when the order was finished. In fact 173.25 tons were actually made and delivered to the Gutta Percha Company. In the Oakamoor wire production book, the Atlantic wire gauge is referred to as '49c', instead of 18, the Birmingham wire gauge figure. This is an illustration of the confusion of gauges which existed at the time, which was not cleared up until the legally enforcable Imperial Gauge was set up in 1884.

Brunel's *Great Eastern* had been hired to lay the cable. She set out from Valentia Bay on 23 July, with a guarantee of £20,000 a year from the Government plus 8 per cent on capital if the project was a success. Over half the cable had been laid when it parted on 11 August, and all the grappling rope was used up in the attempt to recover the lost end. As with the first attempt the cable had to be abandoned, leaving 1,186 miles of the line on the sea bed.

Almost immediately, new cable was ordered. Alfred Bolton broke his holiday in Llandudno to meet Chatterton in Birmingham and take an order for 125 tons of wire. Production began at Oakamoor on 21 September 1865. The wire drawing was finished on 17 May 1866, Boltons having made 113.4 tons of copper conductor.

Instead of simply finishing the existing cable the following year, it was decided to take the opportunity to lay a second cable at the same time. By amalgamating companies, a total capital of £600,000 was raised. Alfred Bolton himself was approached at Oakamoor by Chatterton to take £8,000 worth of shares in the new Anglo-American Telegraph Company. On 30 June, the *Great Eastern* sailed again, this time with one complete new cable, largely unchanged in design from the 1865 one, except for the use of galvanised wire for the armouring, and a length of another sufficient to complete the 1865 cable. On 27 July, Alfred Bolton noted in his diary:

> 1866 Atlantic cable—completed laying 8.45 pm. They have not yet tried to recover the 1865 cable.

This was not done until 2 September, and the old and new ends were spliced at 5.50am, a fact also noted in Alfred's diary. Within a few days both cables were working. A dramatic demonstration of the power of the telegraph to conquer time and space was made by Latimer Clark. He had the ends of the two cables joined at Newfoundland and then sent a strong signal across the Atlantic and back in little more than a second.

The success of the cables was greeted with great public enthusiasm. Many banquets were given in celebration of those who had made the Atlantic telegraph possible. Alfred himself decided to commemorate the occasion personally by ordering a grand piano.

Many more deep-sea cables were laid after the successful outcome of the transatlantic telegraph. The next major concern for Britain was a direct link with India. Parts of this had been attempted before 1866, but with no overall success. Now there was a rush of cable laying to the East up till 1880. Oakamoor was immediately involved, turning out 10 tons of wire for the Abyssinian telegraph in October 1867.

The British were not the only submarine cable layers now. In 1868 the French Government ordered its own transatlantic cable. They turned to Boltons for their copper wire, and Alfred went to Llanelli in August to talk to Nevill about the copper for it. The first wire was drawn on 4 September, and the total order of 216 tons of '56c' wire was finished in April 1869.

The 1870s were busy ones for Thomas Bolton and Sons as the orders for submarine cable wire multiplied. There are no details of these orders, unfortunately. Larger and more powerful plant was installed. In 1870, six new wire blocks were completed. New breaking-down rolls, wire-strip rolls, and large rod rolls were put in. More grooved rolls were being considered in 1873. By 1881, there were new large rolls for rolling wire-bars into coil, and new grooved rolls for rolling rods from the coil ran for the first time in February. By this time the wire-making process in general at Oakamoor looked like this:

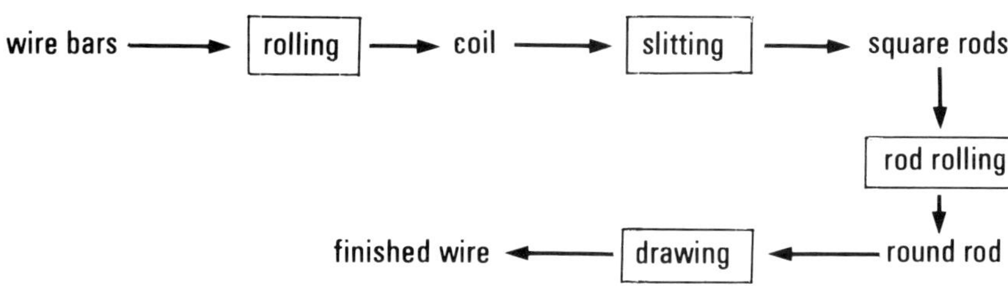

After 1880 the tempo of submarine cable laying dropped, and it became necessary for Boltons to look at the land telegraph for a market for copper wire. This will be part of the story of the next chapter.

On 26 October just after the completion of the Atlantic telegraph, Alfred Bolton received a visit from Professor Holmes about wire for his dynamo for lighthouses. Trinity House had already used Holmes dynamos for the South Foreland light in 1858, and Dungeness in 1862. Now two more machines had been ordered for Souter lighthouse, between Sunderland and Shields. This design had fixed magnets and rotating coils, and the two machines continued in service until 1900. One of them, with wire supplied by Boltons has been preserved in the Science Museum. This was not a large new market: there were only ten electric lighthouses in the world by 1880, of which five were in Britain. But the dynamo was shortly to leave its lighthouse tower and revolutionise the world by the 'mass production' of electricity. This was to open a new chapter in the Bolton's story.

Tubes and Rollers

Most of the Bolton's wire production was transferred to Oakamoor after 1852, and telegraph work was obviously a major interest. But Alfred and Francis did not concentrate solely on that, and restarted the manufacture of two products which the Pattens had made. One was tubing, which Alfred was already familiar with from his time in Birmingham, and the other was copper rollers for printing onto calicoes, or cotton fabrics. This was new to him, but had been part of the Patten's business since 1822. They had acquired the rights to Hollingrake's patent of that year for making printing rollers, in their search for new products and markets during the depression after Waterloo. The equipment used by the Pattens for roller casting was still in place at Oakamoor.

John Collier, the tube foreman was re-instated to take charge of the tube shop, and some manufacture was started. In the first years, however, wire making took precedence at Oakamoor, and developments in the tube-making side were made at Broad Street by Francis Bolton. In 1854 a patent was registered in the names of both of the brothers, although it is clear that Francis had done the work on it. The patent concerned the manufacture of twisted copper and brass tubing, which was in demand for gothic brass foundry, and for gas fittings. These were not really twisted, but were given that appearance by putting spiral fluting on a single copper or brass tube. Francis proposed to do this by drawing the round tube through a die which would impart the fluting to the tube. To make helical flutes rather than straight ones the die was made to revolve as the tube was drawn. To prevent the tube turning as well, Francis proposed to put a short length of rod into the forceps which gripped the end of the tube and drew it through the die. This caught on the bench and prevented the tube from actually being twisted as it was drawn.

The first sign of anything new at Oakamoor came in 1858, when Alfred ordered new draw-benches from Fishers in Birmingham for solid tubing. Seamless tubes, whose manufacture was described in the previous chapter, were mostly used for locomotive and marine boilers. They started life as cast copper cylinders with thick walls which were then drawn down into tubes. Textile printing rollers were also made from cast copper cylinders. But these were usually wider than tube castings, and they had relatively thin walls in relation to their diameter. The rough cast was then turned, ground and polished to make a smooth surface for the engraver to make his design on.

As in the case of the electric telegraph, the Boltons responded quickly to signs of fundamental technical change. Between 1850 and 1900, the Lancashire textile industry began to abandon the old method of block-printing, and go over completely to roller printing, except for a few special applications. The demand for printing rollers was thus on the increase.

Alfred found a way of using the same castings for both solid tubes and for making printing rollers. In February and March 1859 he was busy at Oakamoor putting in equipment for the solid tube shop. By the October of that year, the first orders for rollers were being received, and an agent was being sought in Manchester for them.

Alfred filed a patent, number 68, for the new method in January 1860, which explained the general principles as follows:

> We take a hollow cylinder of copper or alloy of copper, the said cylinder being much smaller in diameter and thicker in substance than it is intended the finished cylinder shall be. After having turned the exterior and bored the interior of the said cylinder, we heat it to redness, and while hot hammer or press it, so as to flatten it. When nearly flattened, we coat the interior with a layer of sand or other powdered solid not affected by heat, or with a mixture of such substance with water or other liquid; we then complete the flattening process. Having again heated it, we pass it between a pair of plain or flat rolls. The flattened cylinder is passed between the rolls according to the method of rolling called 'cross rolling', that is, the axis of the flattened cylinder is parallel with the axis of the rolls. The rolling process is repeated until the width of the flattened cylinder is such that when opened again into a cylindrical form it will have the required diameter. After the completion of the rolling process, the cylinder is opened, and the fins or burrs formed at the places where the metal was doubled are removed. The cylinder may then be finished in the usual or any desired manner.
>
> We sometimes place a sheet of copper or other metal or alloy in the flattened cylinder before rolling it, so as to prevent the sides coming too closely into contact during rolling, and also to prevent too sharp a bend at the edges. The last-named object may also be effected by placing rods or bars of copper having rounded edges at the bends of the flattened cylinder.

Later the same month, Alfred saw Edwards, the patentee of a hydraulic press for making copper rollers. This made him reconsider his own ideas, and for several weeks he thought of using this rather than his own method. Finally in February he gave up the idea, but his ideas were still in flux and he failed to file a full specification within six months as required by law. His own patent thus was void. Nevertheless, Oakamoor began to turn out printing rollers using the 'patent' method, and Alfred also made rollers to other people's designs. A new grinding mill and polishing room were set up to finish the rollers. Demand increased, and Alfred approached the Broughton Copper Co in Manchester, who also made copper and brass tubes, to make rollers for him. In 1866, representatives of one of his customers for rollers, the Burlington Engraving Co, came with an idea for a patent for opening copper tubes. This was an essential part of Alfred's process, and one of the men came back in February to demonstrate the method.

Another example of Alfred's readiness to experiment was the introduction of a water turbine to power the new roller-mill. These began to be used in a small way in the nineteenth century as a more compact source of rotative motion than waterwheels. Various designs were available in the 1860s, and Alfred chose a Schiele impulse turbine. One of these had been on show at the 1862 Exhibition, made by North Moor Foundry Co of Oldham. The machine shop at Oakamoor was cleared so that a hole could be dug in the floor to house the turbine. Wrought iron pipes were laid to bring water to it, and it was started on 4 December 1866. It seemed to run well, but it was found not to be as powerful as had been claimed. A new one was fitted in the following June. The turbine had to be examined again in 1869, but there is no further mention of it, and it probably continued to run satisfactorily for some time. By the end of the century, however, Oakamoor was still drawing its water power from two large waterwheels.

Alfred continued to experiment with new methods of making printing rollers. In September 1871, he tried rolling a smaller tube cylinder into a larger printing roller. This he did by putting it on a mandrel between two flat rolls. The two rolls were driven by pinions, so that they revolved in opposite directions, unlike the normal rolls on a rolling mill. The copper tube was then forced by the mandrel into the gap between the two rollers with its axis parallel to them. He found that the copper cracked badly and broke, but he persisted and got new 22in roller rolls installed by December, which were driven by steam. The following March Alfred was in Manchester discussing machinery for making thin rollers. Part of the original aim of the patent in 1860 had been to make these as well as large-diameter rollers.

More tube-making equipment was bought in the 1870s, starting with some second-hand machinery from Newton Keates works at Holywell which was for sale. Alfred also acquired their tube foreman Swinsin. A new tube soldering shop was set up in 1872 and a tube-turning lathe the following year. By 1876, new tube benches were being planned for the old wire mill, and a new annealing furnace was finished for Swinsin, the tube maker in 1879. The following year Alfred's attention was drawn to Walker's new patent copper tube, and in May 1880 he visited the Patent Tube works in Smethwick in Birmingham.

By 1881, under Alfred Bolton's direct management, Oakamoor had become the most important factory of its kind in Britain. By then it employed between 150 and 200 men. Its products included telegraph wire, tubes for locomotive boilers, brazed copper tubes and brazing materials, copper rollers used in calico printing, and sheet brass and copper of all kinds. Broad Street remained the head office of the firm, but was now comparatively small. It made chiefly brass wire and rolled brass.

One of its specialities was brass wire for the makers of wirewebs for paper mills, and another was dial copper for the watch trade. Broad Street was thus still producing brass and copper fabrications typical of the traditional local industry, and had changed very little under Francis Bolton's management. After his early interest in technical developments shown in his tube patent of 1854, Francis began to spread his business interests outside the firm. He did not have the scientific bent of his elder brother, and at the Proprietary School in Edgbaston to which both he and his brother went, Francis was regarded as 'an exceedingly dull boy'. His business enterprise talents lay more in financial speculations, which he engaged in extensively. Some were successful, some not, including the Midland Land Corporation which he became involved with about this time. This came to a 'disastrous close' in 1884, and Francis Bolton, as one of the directors, lost a great deal of money by it. Another property speculation was the acquisition of a 99-year lease on a prime site in Birmingham's city centre which embraced part of New Street, Stephenson Street, Lower Temple Street, and included many shops and the Theatre Royal. Francis won the tenancy of the property with a bid of £8,000 rental. Francis, then, was not a technical innovator like his brother, and was content to allow the Birmingham factory to continue in its traditional way. It was again to be Alfred who guided the firm into the next phrase of its development between 1880 and the end of the century, which forms the subject of the next chapter.

CHAPTER FOUR

The Electrical Revolution 1880-1902

Steam power had been the driving force of the first Industrial Revolution. Now the second man-made source of power was about to flex its muscles. In the last twenty years of the nineteenth century the mass production of electricity created an unprecedented new source of heat, light and power, made possible by developments in the generation of current from dynamos. The invention of the incandescent lamp by Edison and Swan led to extensive public supply systems rivalling those of the gas companies, and requiring large central generating stations. Electricity supplied a new source of power for traction, both on railways and tramways, and above all power for industry. The ability to produce quantities of electric current made possible the reduction and refining of metals on a large scale by electrolytic methods. The high speed telegraph and the telephone developed rapidly in these years.

All this involved the use of copper on a larger and larger scale, and Thomas Bolton and Sons were in the forefront, making important contributions to the pioneering of all these developments.

Widnes

A supply of high conductivity copper was vital to the Bolton's success in the electrical field, as has been pointed out. So long as Alfred Bolton was dependent on outside smelting companies, in the last resort quality was outside his control. With the development of the high-speed telegraph, the need for high conductivity copper became even more critical. The only solution which would guarantee purity within the limits of the available techniques was to make the copper himself.

The purchase of a copper works had been considered before. As long ago as 1863, while preparing to make the Persian Gulf conductor with his patent rod wire, Alfred had visited the Ravenhead works of Bibby and Sons which they were giving up. While in St Helens he also went round the Pocket Nook works of James Radley. Nothing came of these visits, however.

One problem which may have deterred him was the financial risks involved in smelting, which are not experienced so much by a

Making firebox plates for steam locomotives with a steam hammer at Bolton's Widnes works in 1901.

manufacturer of ready-made copper. The smelter had to buy and keep on hand large stocks of ore, which tied up a considerable portion of his capital. The purchase of ore was always the largest expense in smelting and in addition carried the possibility of losses due to changes in the price of copper during the months that the stocks were held. The position of the British smelters was now quite different from what it had been in Thomas Bolton's day, when most of the copper ore came from British mines. Now the main supply was under the control of the big new American producers in Michigan, Montana and elsewhere.

Nevertheless it had become necessary to go into the smelting and refining business. Up till 1881, high conductivity wire bars were mainly bought from E. Logan of Birkenhead, who were known to refine them from picked brands of 'Chili bars'. So when J. F. Allen of the bankrupt Mersey Copper Works in Widnes called on Alfred Bolton in June 1881 about selling him his works, he found a ready ear. Alfred visited the works a few days later, and bought them on 4 July. At that time they consisted of four refining furnaces, capable of making about 150 tons of copper a month. The intention was to greatly increase the capacity of the works to supply Oakamoor with vital raw material in the form of high-conductivity wire bars, coils of wire rod, ingots and sheet copper. By March the following year the six extra furnaces which Alfred planned were nearly finished, and he found it necessary to acquire more land from neighbouring premises, Robinson's Red Lead works and the Atlas works. A smelting shed was started in May 1882, so that Boltons could choose their own ore, and not only refine the impure 'Chili bars' which was the usual form in which copper was imported. Alfred wanted to ensure that the bars or the ores which he bought contained as few alloys of other metals as possible. In this way, as he pointed out to a meeting of the Institution of Civil Engineers in December 1883, not much oxidising was required in the refining process. Oxides of copper, had of course been known since 1861 to be the worst enemy of the electrical engineer seeking maximum conductivity. There are no details of the precise methods used at Widnes at this time, but W.H.R. Preece, speaking at the same meeting, commented on the very pure copper of up to 99 per cent conductivity produced in the past year or so 'by improvements in smelting and refining furnaces'. This may well have been a reference to Widnes, as Preece, the Post Office's chief telegraph engineer, had met and discussed with Alfred Bolton earlier in 1883.

In 1884 a new refinery was begun, whose construction was supervised by Thomas Bolton, Alfred's elder son. The chimney which he designed

Rolling copper firebox plates on one of the large rolling mills at Thomas Bolton's Widnes works.

Hand ladling copper ingots in the refinery at Widnes, 1897.

and erected was the largest in Widnes at that time. By then several rows of houses and two streets had been absorbed, as had St Patricks, a Catholic church in Hutchinson Street. A new church was built in Dock Street, Widnes, partly with the compensation paid by Boltons. Later the Catholic Schools were acquired for use as offices. The expanding Mersey Copper Works was heard as well as seen, through the works hooter. This came from the *Great Eastern*, which had carried the cables in 1865 and 1866. It was used to announce the Relief of Mafeking, the end of the Boer War and peace in 1918.

As well as smelting and refining, Alfred from the outset set up a rolling mill. It was Richard Nevill's men from Llanelli who were called in to do the work of erecting the equipment. This steam-powered mill was ready by the summer of 1883. By 1885, rolls capable of taking plates 11ft wide had been installed. Widnes was no longer merely supplying Oakamoor but had become an important source of copper wire bars for the trade, and of copper plates and stay-bolts for the fire boxes of locomotives for the railway companies and other locomotive makers.

To cope with this considerable expansion in the activity of Thomas Bolton and Sons, the partnership was expanded in 1881. In addition to Alfred and Francis Bolton, the firm was to include Thomas Bolton, Charles Bolton Toller and George Rathbone. Thomas had been educated at Harrow from 1871 to 1874, and then like his father went to University College in London, where he stayed for three years, studying under Sir Alexander Kennedy in his engineering laboratory. After that he went to Kitsons the locomotive manufacturers of Leeds for an apprenticeship on the shop floor, but again like his father, he had to return to the business sooner than intended because of Alfred's ill-health. He started work at Oakamoor in 1878. Being the eldest son, and also because he shared his father's enthusiasm for scientific and technical matters, he was to be groomed as his father's successor. He was later to be responsible for the continuous wire-drawing machine, and other innovations developed in the firm's works.

Charles Bolton Toller was a son-in-law of Alfred Bolton, who had married his second daughter Alice in September 1877. He started to work for the firm in 1875, and was put in charge of the Widnes works in December 1885. Alfred decided on 'Charlie' rather than Thomas because he wanted to keep his son at Oakamoor which was still the centre of the business. Charles thus became the firm's chief buyer of copper, which gave him the largest budget of all the partners.

George Rathbone was a member of the well-known Liverpool merchant family, which had become wealthy through the cotton and tea trade. Thomas Bolton had married his sister Nina Rathbone in November 1880, so all the partners had family as well as business ties. George had been working for the Boltons since 1876, when he was sent from Liverpool to the Broad Street Office in Birmingham. He was later

Some of the Bolton family at Moor Court about 1890. As may be seen they were all keen cyclists.

given a one third share of the profits from Birmingham in addition to his one-ninth of the general business. It is also possible that the Rathbone connection brought new capital into the firm which helped pay for the expansion at Widnes. However, Alfred himself was becoming a wealthy man by this time. In 1872 he had paid out sums of £20,000 and £15,000 to buy Cheadle Park and Litley farms, and Sweet Hills plantation. The first also brought him the manorial rights of the manor of Cheadle. But Alfred was more interested in the mineral potential of the area than in getting himself into Burke's *Landed Gentry*, and he spent a considerable sum later in opening the Cheadle Park colliery on his own land. The first coal from this mine was brought up in June 1890, and two large pieces mounted on a wagon were paraded round Cheadle before being taken up to Moor Court, the house which Alfred had built for himself overlooking Oakamoor.

Electrolytic Refining at Widnes

The next big step taken by Thomas Bolton and Sons was to go over completely to electrolytic methods for refining copper for electrical uses. Conventional fire refining of copper had reached the limit of its effectiveness at Widnes. Alfred Bolton was able to exhibit samples of copper with a conductivity of 99 per cent and more in 1883. But the new developments in high speed telegraphy and elsewhere demanded higher purities, and this made electro-depositing a necessity for Boltons.

The basic process involved submerging an anode of impure copper and a cathode of pure copper in a vat of copper sulphate. The passage of an electric current through the solution from anode to cathode transfers pure copper to the cathode and leaves any impurities in the sludge at the bottom of the vat. Electrolytic methods were not of course new. Silver plate had been made by electro-depositing silver onto copper under Elkington's patent since 1842. James Balleny Elkington pioneered the refining of copper by electrolytic methods in 1865, and recovered the gold and silver in the impure metal from the slimes left after the depositing process. The method was first used at the Pembrey works in South Wales in 1869 by Mason and Elkington. Similar machines were later installed by Vivian at the Hafod works and by Williams Foster at Morfa, both in South Wales. The patent method envisaged magneto-electric machines with fifty permanent magnets weighing 28lb each.

Large-scale application of electricity to the refining of copper had to wait for developments in the generation of electric current. These began with the discovery of the dynamo proper in 1867. Up till then all generating machines, including Elkingtons, had been 'magneto-electric', that is they used the magnetic force of permanent magnets to create electricity. Almost simultaneously in 1867, Wheatstone and Varley, on the one hand and Siemens on the other showed that permanent magnets were not necessary for converting mechanical movement into electricity.

'Electro magnetism', or magnets augmented by electricity produced proportionally very much larger amounts of 'dynamo electricity'. Hence the 'dynamo-electric machine' came to replace the magneto-electric one.

At first, the new dynamos suffered from unsteady current, which made them unsuitable for most purposes. This was overcome by the French engineer Zénobe-Théophile Gramme. In 1870 he patented the Gramme dynamo with its ring armature, which opened up a new era in electrical engineering. By winding many separate coils of copper wire on a ring of iron, he was able to produce a steady current, and maintain it over a long period without overheating the dynamo. He built his own factory to exploit the patent and it was not till this had expired in 1884 that others were able to develop it further.

Alfred Bolton first became interested in electro-deposition of copper in 1883. It was said that W.H. Preece first asked him to look into electrolytic methods for making telegraph wire. While in London in November 1883, Alfred met Partz, a German experienced in depositing metals electrically. He told Alfred that he should use dynamos that had the least possible internal electrical resistance for successful deposition. The following day at a colleague's house, Alfred had a 'long talk' with a Spanish marquis who owned a copper works at Santa Constanza, where he refined his copper by the electrical process.

By 1885, Alfred had become convinced of the necessity for installing electrolytic refining plant at Widnes. One of the first men he approached was Professor Forbes who had a new dynamo based on Gramme's ideas. But in the end Alfred decided to use Siemen's dynamos, since they came from an established manufacturer. Building began in 1885, and the lead vats in the depositing sheds were finished by the end of January 1886. The plant was thought to be ready for depositing to start in July, and Niebel, a representative from Siemens was present for the first attempt on 14 July. This had to be delayed because the vats were not yet full of the copper sulphate solution. When the first trial was made on 23 July, the vats were found to be no good. Niebel explained that this was because they had been built into the ground. The introduction of electro-depositing was thus delayed for another year.

This proved fortunate, because it allowed Alfred Bolton to make use of the latest advance in dynamo design, which had been made by Thomas Parker of Wolverhampton. Parker had set up there in 1882 in partnership with Bedford Elwell. He developed the electrical side of the business, making accumulators incorporating Parker's discoveries. He provided dynamos and lighting equipment in 1883 for a Forest of Dean Colliery, supposedly the first in the country to be lit by electricity, and in 1884 initiated the Blackpool tramway system.

After 1884 and the expiry of Gramme's patent, he developed a more efficient and cheaper dynamo, and in May 1887 came to Oakamoor to talk to Alfred Bolton about using them at Widnes. Alfred was convinced and ordered two 1,500-amp machines, a decision he was not to regret. So enthusiastic was he about Parker's dynamos that he used them to light the

The Elwell-Parker dynamo sets at Widnes, 1897.

The copper electro-deposition tanks at Widnes showing the different levels.

works and consulted him about lighting Alton Towers and the village of Farley as well.

The use of Parker's dynamos at Widnes from 1887 was said by W.H. Preece to have 'revolutionised the process of purification' of copper. These special dynamos were capable of depositing 18 tons of copper a week, and by the mid-1890s there were four Elwell-Parker dynamos working at Widnes, making about 300 tons of electrolytic copper a month. The dynamos ran continuously for many years and gave very little trouble.

The first error with the tanks was corrected and new ones were built on a series of levels, so that the liquid could run down from one tank to the next and be in constant movement.

Electrolytic refining was more expensive than the conventional reverberatory furnace method. To precipitate the pure copper from the anode might take up to two months. This meant that a considerable stock of copper must always remain in the depositing vats, and its cost would have to be borne by the refiner. For this reason, electro-depositing was only economic if the precious metals, mainly silver, were recovered from the slimes at the bottom of the vats. Many of the imported coarse copper pigs and ores contained precious metals as 'impurities', particularly those from the Anaconda mine in the USA. This was not charged for at first, and so the British refiners could reap the benefit. Eventually the American producers stopped sending silver-rich ores, and this forced the closure of the electrolytic refinery at Widnes, as it was no longer economic.

For the time being however, all was well. The 'argentiferous pigs' and ores were received at Widnes and melted down in the furnaces. The copper intended for electrical use was poured into flat moulds to make anodes for the depositing vats. The pure cathode copper was then re-melted to make wire bars and high-conductivity copper ingots for Oakamoor and the trade. The sludge was taken to a silver refining shop, which was set up at Widnes in December 1887.

By the mid-1890s, the Widnes site covered four acres, and was producing 1,000 tons of copper a month, of which about 300 tons was electrolytically refined. The rest went into making large quantities of locomotive fire-box plates and bolts, and other plates of all shapes and sizes up to 12ft wide. The future expansion of electro-refining however was to take place elsewhere.

The Secretan Episode

Thomas Bolton and Sons were now one of the major buyers and users of copper in Britain. As a result of their success they were involved in the first of the many attempts to manipulate the price of copper. This was the audacious Secretan syndicate, which attempted to establish a 'corner' in the world copper market between October 1887 and its collapse in March

1889. It was essentially a combination of French finance with the USA copper producers, who were by then supplying 40 per cent of the world's output. M. Secretan was a director of the Société Industrielle et Commercialle des Metaux, Europe's largest copper manufacturing company. Backed by money from French banks, he offered three-year contracts to the world's main producers, particularly the Anaconda group, and the Calumet and Hecla group in Michigan. The Americans were to have control of the sales at home, and the French operators of the syndicate were to market the copper in the rest of the world. The immediate result was that the copper price doubled from a rather low figure of £42 a ton to £82 a ton. By the spring of 1888, the British were being forced to consider terms for joining the syndicate. On 19 March, Alfred Bolton and Charles Toller, his buyer, met with Schwann, a leading copper dealer, H. Druce, and Grenfell at the London office of the other major copper smelter, Vivian's, to discuss the matter. They met again in April, this time also including leading Birmingham industrialists Muntz and Everett. They met M. Secretan for the first time at a dinner at the Carlton Club, and the next day Alfred was at a meeting of copper tube makers who discussed joining the syndicate.

The 'corner' eventually collapsed in March 1889, because it proved impossible to really control the supply of copper from sources outside the USA. M Secretan committed suicide rather than face the consequences of the debacle. For Bolton's however the outcome was more favourable. Since opening the Widnes works they had become suppliers of copper to others in the trade and they stood to benefit from membership of the syndicate and enforced higher prices. It was said later that they made enough from the Secretan operation to pay for their purchase of Widnes and the expansion they had undertaken there. This no doubt did not square with the Liberal politics in which Alfred was becoming more and more involved. He was elected a member of the National Liberal Club and was elected as a Liberal councillor in the first Staffordshire County Council elections in 1889, with 599 votes as against 518 for the Conservative. However, the principles of Free Trade did not inhibit him from accepting the profits of this exercise in absolute monopoly.

Copper Telegraph Wire

Until the 1880s the large market for overhead telegraph wire had gone almost entirely to the iron wire makers. Now for the first time copper wire was penetrating this market and was soon to conquer it entirely. The reason for this was the development of hard-drawn copper wire which combined good tensile strength and superior conductivity. This breakthrough was initially achieved in the USA by Doolittle. By using special drawing plates, he had produced a high conductivity rod which could be drawn into wire without heat treatment, and thus doubled its tensile strength. The development of high-speed telegraph, and the

The flywheel of one of the steam engines at Oakamoor.

Another steam engine at Oakamoor.

telephone, had revived interest in the superior electrical qualities of copper wire.

In Britain it was Alfred Bolton who pioneered the production of hard-drawn copper wire at Oakamoor, where he made telegraph wire with a breaking strain of 340lb, which he considered to be sufficient for all purposes. Alfred's commitment to hard-drawn copper as the best alternative to iron meant that a great premium was put on purity and conductivity. Copper wire that was hard-drawn, or unannealed, lost up to 2 per cent of its conductivity. So it was essential to start with the very purest copper, which only electrolytic refining could provide. This lay behind Alfred's decision to go for this method of refining.

In Europe, by contrast, different solutions were found to the problem of making copper wire suitable for overhead use. Where Alfred Bolton openly rejected the use of additives to his copper, and declared that experiments in his works had shown that adding substances to copper led to a loss of conductivity, Lazare Weiller in France had taken the opposite course. In 1882, he took out a patent for silicon bronze, a copper alloy which contained 2-4 per cent silicon, as well as smaller amounts of tin. This was shown to have very superior tensile qualities, as well as good conductivity. It was used for telegraph and telephone lines between Paris and Brussels, and later for power cables to light the boulevards for the 1889 Paris Exhibition.

Alfred Bolton found a powerful ally and advocate of copper wire as against iron wire in W.H. Preece, the chief telegraph engineer of the Post Office. This association with Preece which proved mutually beneficial to both parties, had begun in November 1883, when the two men met at the Wimbledon home of Schwann, the copper dealer. On 9 December they were together again at a meeting of the Institution of Civil Engineers where Preece was giving a paper on electrical conductors, in which he spoke up for copper as an overhead conductor, and mentioned favourably the virtues of the new hard-drawn copper wire. Alfred was present with samples of his wire to show the distinguished audience, which contained many prominent people in the electrical and telegraphic field.

Immediately after this meeting, users of overhead telegraph wire began to beat a path to Alfred Bolton's door at Oakamoor. In January 1884, Randall, the telegraph engineer of the London, Chatham and Dover Railway Co, called about the new wire. In March, Alfred met Spagnolette, the telegraph engineer of the Great Western Railway. In the course of the year, he saw the two main telephone companies, the United and the National, the Caledonian Railway telegraph engineer, and the L & D Electric Wire Company. He also visited a Captain Addison in Whitehall about military uses for the hardened copper wire.

The following year Alfred put in a tender for supplying telegraph wire to the War Office. He supplied wire for the Suakim Railway telegraph, and had a meeting with Fletcher, a representative of the United Telegraph Co. In 1886, the London and South Eastern Railway telegraph engineer showed interest in the hard-drawn wire.

But the most important customer of all was the Post Office. In January

1885, Alfred called on William Preece in London with samples of his wire for testing, drawn to various levels of hardness from electrolytic copper. These were tested by J. Bell at the Gloster Road Postal Telegraph factory for durability and strength over the next month. On 28 March, Ford the Post Office inspector came to inspect wire prepared for him from the Oakamoor stores. In July and the following January there were further tests carried out by Roberts. The outcome of these visits was a new specification for hard-drawn copper wire supplied to the Post Office. Wire was to be graded in terms of weight — so many pounds of high conductivity copper per mile. The terms of the specification show the rigorous nature of the tests, and the high standards which Boltons were expected to maintain.

1 The wire shall be drawn in continuous pieces of the respective weights and diameters given in the table and every piece may be gauged for diameter in one or more places.

2 The wire shall be perfectly cylindrical, uniform in quality, pliable, free from scale, inequalities, flaws, splits and other defects, and shall be subject to the tests hereinafter provided for.

3 Every piece may be tested for ductility and tensile strength and 5 per cent of the entire number of the pieces may be cut and tested in any part. Pieces cut for this purpose shall not be brazed or otherwise jointed together, but each length shall be bound up into a separate coil.

4 The wire shall be capable of being wrapped in six turns round wire of its own diameter, unwrapped, and again wrapped in six turns round wire of its own diameter in the same direction as the first wrapping, without breaking; and shall also be capable of bearing the number of twists set down in the table without breaking. The twist test will be made as follows:- The wire will be gripped by two vices, one of which will be made to revolve at a speed not exceeding one revolution a second. The twists thus given to the wire will be reckoned by means of an ink mark which forms a spiral on the wire during torsion, the full number of twists to be visible between the vices.

This twist test was observed at Oakamoor some ten years later by the deputy editor of the *Electrical Review*. Where the Post Office stipulated twenty-five twists should be visible, the wire did not break until it had been twisted sixty times. This was a measure of how the special relationship which developed through Preece with the Post Office worked to the benefit of both customer and supplier. The first got wire of the highest standard, while the firm had to improve its methods and equipment to maintain this level of quality.

The first fruit of the association was the London to Dublin telegraph line for which Boltons supplied 150lb per mile hard-drawn copper wire. In December 1886, Alfred Bolton, W. H. Preece, and three other Post Office engineers, Gary from Cardiff, Evans from Birmingham, and Edwards from Liverpool, went to see the new line being erected near Moreton-in-the-Marsh. Earlier that year, appropriately, Alfred was elected to membership of the Society of Telegraph Engineers and Electricians, the forerunner of the Institution of Electrical Engineers.

A narrow strip rolling mill at Oakamoor.

An early strip rolling mill at Oakamoor in 1946.

That the professional organisation developed in this way, shows the importance of the telegraph in the general development of electricity. The change of name was in fact agreed on the year after Alfred joined.

In the years which followed Oakamoor became the principal source of overhead telegraph wire for the Post Office, and the India Office, which ordered 120 tons of wire in 1887 and another 276 tons in 1893.

Although the telephone companies were interested in and used Bolton's hard drawn copper wire in the 1880s and early 1890s, this did not prove such a large market. The reason for this was the slow development of the telephone in Britain, and for this W. H. Preece may be said to be partly responsible, for the Post Office was jealous of the new form of communication, fearing a threat to its established telegraph links. The telephone had been introduced into Britain in 1877 by Alexander Bell, but by 1885 there were still only 13,000 telephones here compared with 300,000 in the USA. Europe was ahead of Britain too, having 17,000 kilometres of long distance telephone line by 1887, whereas London was not connected to any other town except Brighton till 1890. Demand for copper telephone wire was not significant till the 1890s. In 1893, Preece placed an order for a large quantity. But by this time Oakamoor hard-drawn wire had met a powerful rival in the silicon bronze wire patented by Weiller in 1882. At first bronze wire had been too brittle but this problem had been overcome. In the end silicon bronze was to replace hard-drawn copper for all overhead conducting purposes.

This challenge had serious commercial consequences for Thomas Bolton and Sons in the field of telephone wire in particular. In 1893, Weiller set up a large factory near Angoulême in France to make bronze telephone wire specifically for the English market. Free trade meant of course that there was no restriction on the entry of his products. By the time-honoured method of dumping, Weiller wanted to force Thomas Bolton and Sons into a ring with the other European manufacturers for sales of telephone wire in English markets. So successful was he that Alfred was forced to get in touch with Weiller, who came to Moor Court in November 1894. To prevent a price cutting war, Alfred Bolton agreed to take Weiller's wire to stop him selling in England. In December Thomas Bolton was sent to Cologne, the home of the largest European wire makers, Felton and Guilleaume to discuss a wire association. The terms of the deal were very onerous for Boltons. By the terms of the agreement drawn up and signed on 28 November 1894, the firm guaranteed to buy up to 200 tons of Weiller's wire and rods every month. By 1896, the burden of these enforced purchases was becoming too much and Alfred tried to get the courts to annul the agreement. The first judgement, handed down on 7 February 1886 was rather indefinite. It declared that Boltons were bound to take Weiller's wire and rods; and on the other hand Weiller was bound to make concessions to enable him to take the copper and still be able to meet competition from America and elsewhere. The basis of this second condition was not one of the terms of the original agreement, but a statement made in a letter from Weiller.

Datestone T B & S 1885, on the hard wire mill, Oakamoor.

The latter was not satisfied with this and on appeal won damages against Boltons on the grounds that his letter did not constitute a binding contract. The family were naturally furious with their counsel, Sir Edward Clarke, who they thought had made a mess of the defence. Despite the loss of the court case, the episode revealed the dominance which Thomas Bolton and Sons had over the domestic market for overhead wire. At the same time, it also showed that the firm would now have to fight to keep ahead of competition both at home and abroad if it was to maintain its position.

The New Wire-Bar Rolling Mill

The greatly increased demand for telegraph wire from Oakamoor meant changes in the methods of production. The old method of rolling and slitting the copper to get wire rods was too slow to match the output required. The same problem had faced the iron wire makers earlier in the history of the telegraph, and they solved it by using the Bedson process of rolling wire rods directly from wire bars (billets). W.H. Preece noted that telegraph engineers owed a great deal to George Bedson. Alfred Bolton was now faced with the same problem, and found the same solution.

The first rod mill for rolling copper wire rods directly from wire-bars was set up at Oakamoor in 1887. Thomas Bolton was given the responsibility for its erection. A start was made in February, and Thomas paid a visit to Lord Dudley's Round Oak Iron Works at Brierley Hill to see for himself iron wire rods being rolled from billets. The machinery for

The first rod-rolling mill at Oakamoor, started in 1887. The breaking-down rolls are in the centre, with the finishing rolls in the background on the right.

the new mill at Oakamoor, made by Taylor and Farley, was installed in May and in August the new engine for it was tried out. By 11 November, the new mill was at work. For the first time, non-ferrous metal wire was made in Britain without the need to roll and slit the metal first. Some wire was still made by the old method at Oakamoor, for special operations. But the new methods revolutionised the manufacture of long lengths of copper wire. Now the only limitation on the length of the wire rods, and hence on the length of the wire itself, was the size of the wire bars which the rolling mill could handle. Boltons were not to rest content with their 1887 breakthrough.

Continuous Wire Drawing

As electrical uses for wire expanded, the need developed for fine wire of around 35 gauge or below, which was about 15 gauges smaller than the standard wire. Brass wire makers were also interested in fine wire. Experiments began on machines that could produce wire of this kind. It soon became clear that the only way that standard wire could conveniently be drawn down to such fine dimensions was by continuous drawing. This does not mean that the drawing goes on until the machine

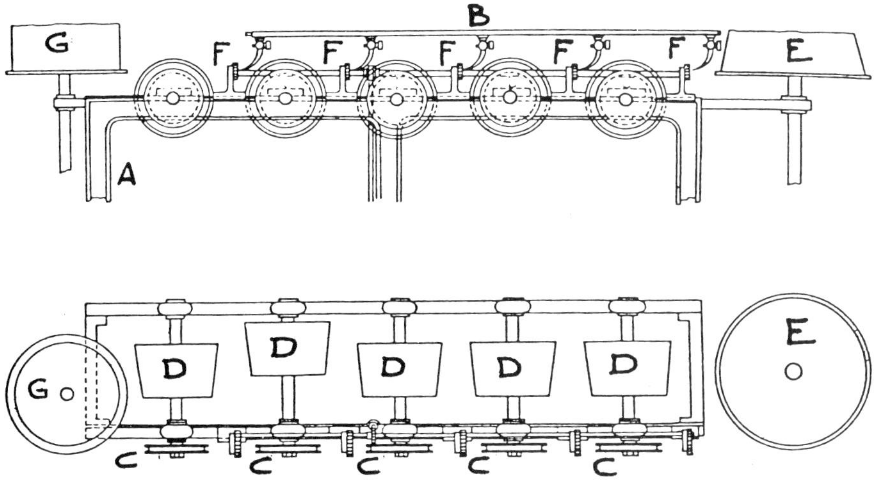

Key:
A The draw bench.
B The lubricating system.
C The wheels round which the wire is lapped before it passes to the next die.
D The drums driving the wheels C.
E The drawing block which draws the wire through the dies.
F The dies set along the bench.
G The reel from which the ordinary wire is unwound.

The S.H. Byrne continuous wire-drawing machine. A diagram from his patent of 1885.

is stopped, but that the wire is continuously reduced in size as it passes through a succession of dies. The main problem with this is that the wire must not be under tension on both sides of the die, otherwise it will snap. With a single die this difficulty does not arise. But when a wire is being drawn from one die through another, there will normally be tension on both sides of the second die.

One of the first men to overcome this problem was S. H. Byrne of Ramsden and Camm's of Brighouse, near Huddersfield, who patented his continuous drawing machine in 1885. He set his dies in line along the draw-bench. The wire was driven between each of the dies by passing it over a drum. Each of these drums was driven at different speeds to accommodate the increasing speed and length of the wire as it passed from one die to the next. This involved a complex system of gearing to adjust the speeds of ten or more different drums. In addition, the dies were also made to revolve, and were kept lubricated by a continuous supply of soft-soap solution fed through pipes. The machine worked, but it had at least two disadvantages. One was the complexity of its mechanism, and the other was that its inventor sold the patent to foreign manufacturer.

Alfred Bolton had become interested in developing a fine-wire

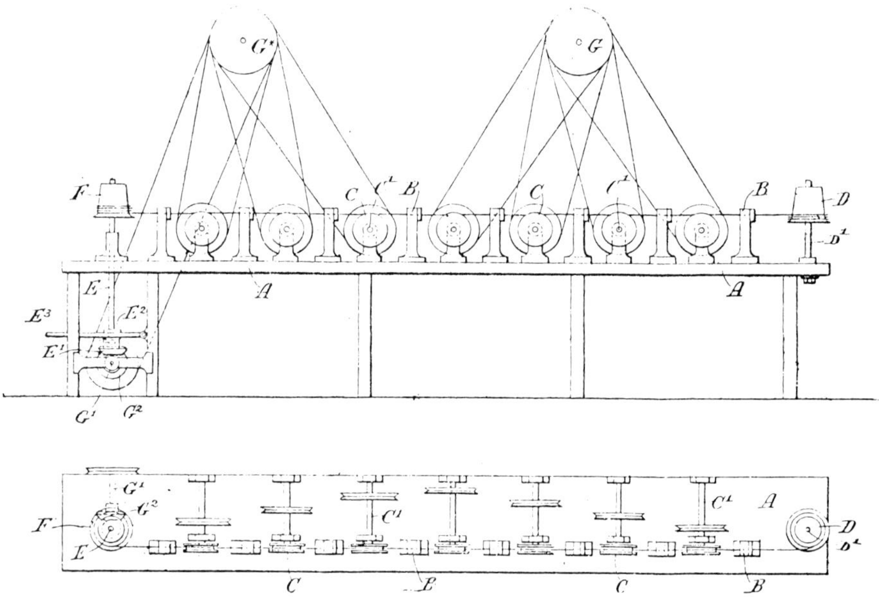

A.S. and T. Bolton's first continuous wire-drawing machine from the 1886 patent drawings. The basic similarity to Byrne's machine is clear.

machine and went to Brighouse with his brother and George Rathbone in January 1886. They saw the Byrne machine at work drawing down wire to fine sizes. But they also took note of its disadvantages, and Alfred and Thomas Bolton set themselves the task of improving on Byrne's solution. Within a few months they had developed a machine which they patented provisionally in July 1886. Patent No 8977 looked very much like Byrne's: the dies were laid out along the length of the draw-bench, and the wire passed over drums between each die. The difference was that the drums all revolved at the same speed, which was in excess of the fastest speed demanded by the rate of travel of the wire through the dies. This meant that in all cases the drums would be revolving faster than the wire. The idea was that the friction of contact between the drum and the wire lapped round it would be enough to ensure adequate draught through the previous die, and at the same time ensure that the wire would not be under tension as it passed into the next die. By having the drums all revolving at the same speed, Alfred and Thomas Bolton simplified the machinery needed. The friction method worked well, and in July 1886, a new fine wire shop at Oakamoor was started to house the new machines.

The final specification for the patent was filed in April 1887. But Alfred and Thomas had become dissatisfied with the improvement they had made. Their machine, like Byrne's, still suffered from the other weakness that the dies were arranged lengthwise along the bench. This would take up a lot of space if say fifteen dies were needed. It also lacked flexibility, if

End view

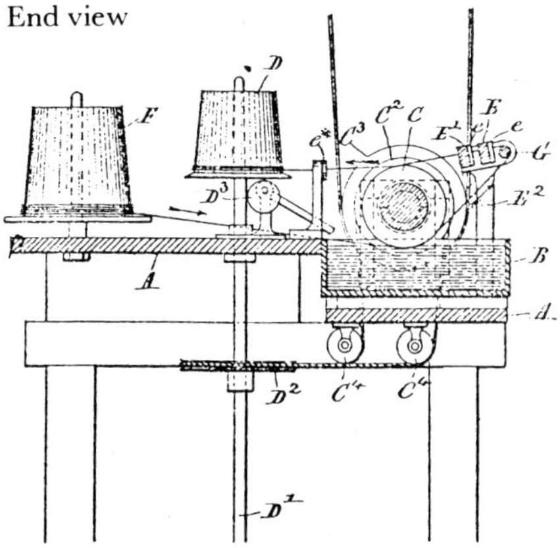

Key:

A	The draw bench.
B	The bath of lubricating fluid for the barrel C.
C	The barrel or drum driven by the belt drive shown. The wire is lapped round the drum and then returned to the dies via the guide rod G.
D	The block which draws the wire through the succession of dies and collects the finished fine wire.
E^1,E^2	The dies, set in pairs of equal size.
F	The reel from which the standard wire is unwound.
G	The guide rod.

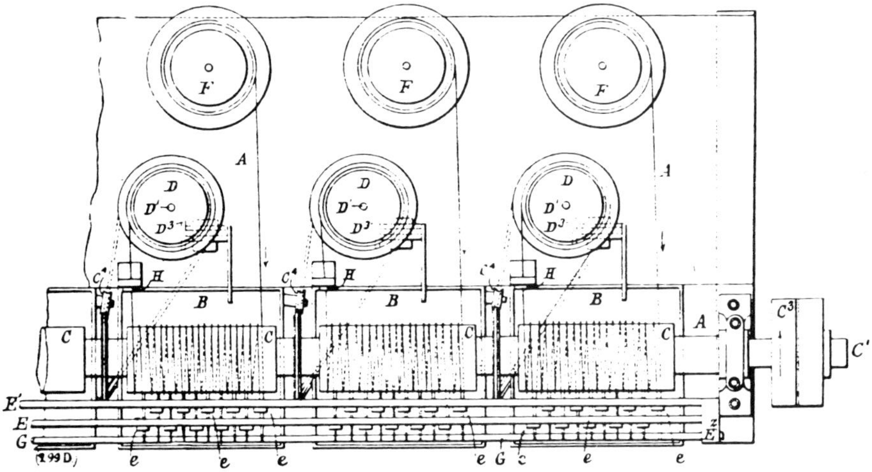

Bolton's second continuous wire-drawing machine, patented in 1887. The plan shows the very compact arrangement of dies and drawing blocks which the design makes possible.

the number of dies was to be often changed, as different orders required. The friction-drive principle was simple and effective and was retained, but the rest of the machine was redesigned. Thomas made some drawings for the new fine-wire machine in March 1887. The old specification however went ahead and a fine-wire machine of the original design was run for the first time in May 1887. The next day, 18 May, Byrne came from Brighouse to see it easily draw down a piece of standard wire. But by 25 May, Thomas Bolton and his younger brother Frank were already trying a new experimental wire-drawing apparatus.

This machine involved a simple extension of the principle of the first. If all the drums were revolving at the same speed, then there was no need for

An early tandem continuous wire-drawing machine, with up to twenty-one dies.

Continuous copper wire-drawing at Oakamoor.

An early continuous wire-drawing machine. One of Thomas Bolton's original machines was still in use at Oakamoor in 1934.

separate drums for every die. One long drum or barrel would do as well. The dies could then be set together in a row facing the barrel. This was the design which Thomas Bolton arrived at. The wire was taken from the reel and lapped round the barrel or cylinder. It was then passed round a guide bar, placed parallel to and in front of the rack of dies, and into the first die. From there it was again lapped round the barrel, round the guide bar and then fed to the next die. Instead of a complex lubrication system for each die the barrel itself was simply partly submerged in a suitable lubricant.

Thomas had produced the first really successful fine wire -machine. It was compact and simple, and several benches could be installed in the space taken up by the original design. The patent, No 8133, was taken out in June 1887, and the complete specification accepted in April 1888. The only modification which was later made was the use of a cone-shaped barrel to allow for the greater length of the wire as it got thinner.

The experimental apparatus which was rigged up in the fitting shop at Oakamoor in May 1887 contained ten diamond dies. These became a feature of the machine. Boltons were not the first to use them, but they did become a speciality of the firm because of the continuous wire-drawing machine. A large stock of diamond dies was kept at the works. The use of diamonds ensured a lasting accuracy at the fine dimensions involved.

By 25 August 1887, the apparatus had become a machine proper, and it was started for the first time in a room over the stranding shop. Within a few months, the first of several manufacturers came to inspect the new machine. In November and December Smith and Patchett, both from Halifax, and J. Clark Johnson from Manchester came to see it. In June the following year Boltons showed a fourteen-die machine at the Glasgow Exhibition, which drew 34 gauge (.0092in) wire down to 48 gauge (.0016in). This was followed by a visit from the firm of Johnson Hobbs. In 1889, the machine was shown at the Paris Exhibition and this aroused interest among manufacturers in the USA. Boltons had secured the American patent rights, and in fact it was from there that the strongest interest was shown. In May 1889, the Coe Brass Manufacturing Co, in Torrington, and the ex-patriate Greening Wire Co of Hamilton, Ontario both sent representatives to see the machine, to be followed in August by another visitor from the Scoville Manufacturing Co of Connecticut. In June the first large British order came: sixteen machines for Patchett of Halifax. The transatlantic interest in the machine was one of the reasons for a two-month visit to North America made by Alfred and his son Frank in September and October 1889.

The principles of the continuous wire-drawing machine became universal in later machines for fine wire. Bolton's original machine remained in use almost until World War II at Oakamoor for some special jobs.

Thomas Bolton and Sons and the Supply of Electricity

The most striking aspect of the electrical revolution in these years was the growth of the supply of current for lighting, power and traction. Alfred Bolton and the firm were closely associated with these developments and with two of the leading figures involved, R. E. Crompton, and S. Z. de Ferranti.

Gas lighting was of course already well-established in the home, in the streets and in public buildings. But electricity meant less fire risk, and would be preferred if it was cheaper. The first challenge to gas came in the 1870s from arc lights and 'Jablochkoff candles', which were suitable only for lighting large rooms and buildings or for streets, railway stations, marshalling yards and other similar uses.

The incandescent lamp developed by Swan and Edison simultaneously by 1880 meant that the units of illumination could be smaller, and they were much more suitable for house lighting. Electric light now had universal application. The first successful installation using the new incandescent lamps in public buildings and streets was set up in London at Holborn Viaduct in 1882. In conjunction with the city authorities Edison and Swan established a small power station which supplied the Old Bailey, the Post Office and the City Temple from two 1,000-lamp 'Jumbo' generators. The current was carried by the new 'VB' cables made by Callender's Bitumen Telegraph & Waterproof Co Ltd, which was set up in 1882 to exploit the discovery of vulcanised bitumen. It may well be that the copper in the cables came from Boltons, as the new cable company started to buy Oakamoor copper wire from the beginning, but there is no definite evidence of this.

The Grosvenor Gallery installation is generally regarded as the world's first public supply system. The New Bond Street enterprise began as a private local lighting installation, but developed into the pioneering example of public electricity supply on a large scale. Two small steam engines were used initially to drive two Siemens Bros 2,000 volt alternators. The gallery was lit by arc lamps fed with a steady 10 amp current. The local residents and shop keepers were very impressed and asked to be linked up with the installation. The demand grew to such an extent that a more permanent generating station had to be set up in a large room dug out underneath the gallery. Two Siemens alternators, each of 250kw output were put in, these being the most powerful yet built. A system of overhead lines began to extend further and further across the rooftops to carry the current to the customers. By 1886, the installation was in need of expert management, and the rising young star of the electrical revolution, Sebastian Z. de Ferranti was put in charge.

In the same year, the first house lighting supply system using incandescent lamps was begun in nearby Kensington. The leading figure in this was another future house-hold name in the world of electrical engineering, Rookes Evelyn Crompton. He had begun with a factory at Chelmsford in 1878, making arc-lamps and importing Gramme dynamos. His first major contract was to light the Vienna Opera House in 1883. The gas-lit Imperial Theatre had burnt down, and an electrical system had been advised for safety's sake in the new building. Crompton's installation used direct current to supply 100 volt incandescent lamps. Vertical steam engines specially designed by Willans and Crompton for directly driving dynamos were used for the first time, and the installation was a spectacular success.

After this, Crompton was on the look-out for opportunities to show what he could do in England. The chance came with a newly-developed housing estate in London. The notorious Baron Grant had bought some slum property called Short Rents south of Kensington High Street. He cleared the houses, built a palace for himself, and then went bankrupt. The estate passed to the Land Securities Co, who began to develop it by laying out streets in the cleared property. Crompton persuaded the company to light the estate by electricity, and suggested that he build a small generating station for the purpose. He then persuaded a few men 'of his acquaintance' with an interest in such electrical ventures to put up the £10,000 required for capital. Among these men was Alfred Bolton, 'a leading man in the copper trade' as Crompton recalled in 1930. Other directors of what came to be called the Kensington Court Electric Light Co were Sir Richard Webster, later Lord Chief Justice, Granville Ryder, the chairman of Land Securities, and Sir Frederick Bramwell, the well-known engineer.

The first meeting of the company board, attended by Alfred Bolton, took place on 30 November 1886, at 1 Great George Street, the head office of Land Securities. Crompton arranged with Alfred for him to supply high-conductivity copper strip 1in wide and ¼in thick for the mains. The estate was served by a system of subways 6ft high and 3ft wide, originally intended for hydraulic power mains. The bare strip was fixed to the walls of these subways on porcelain insulating brackets designed by Crompton. These allowed for several ¼in layers to be added, up to 1¼ in thick, as the system expanded. It began at a low pressure of 100 volts, and the connexions were carried into the houses through holes in the subway walls. Ferranti was consulted about meters for the consumers. The generating plant consisted of seven Willans-Crompton sets, delivering 500 amps at 100 volts and was sited near Kensington High Street. A large battery of accumulators was installed to take the load during the night, as was usual with direct current systems. This made them initially more popular than the alternating current systems advocated by Ferranti and others, in which battery power could not be substituted.

The plant was started up in January 1887, and by the end of the year, the Kensington Court Co had begun to show a 5 per cent profit. Neighbouring residential areas wanted to join the system, and a second generating set was put in before the end of 1887. Bolton's bare copper mains were extended along brick culverts in the street, covered by the paving stones. Customers were charged the equivalent of 3p per kwh. In 1888 the company was reorganised as the Kensington and Knightsbridge Electric Light Co, with a capital of £250,000. It had now spread far beyond the confines of the Kensington Court estate. In the same year, the new Lighting Act and the subsequent Marindin Inquiry gave the Kensington and Knightsbridge Co a larger area. The Kensington Court station was enlarged, and another one built at Chapel Place in Knightsbridge in 1890. Over the next decade the company continued to expand, and Alfred Bolton was a constant attender at the board meetings. By the end of the century, the demand for current was so great that the company found it necessary to follow the example already given by Ferranti and set up a large generating station outside the urban area to supply their customers. This led to the establishing of the power station at Wood Lane, Shepherd's Bush in 1900.

The Boltons first acquaintance with Ferranti was made in 1883. Then only 18 years old, he visited the Bolton family while they were spending Easter at Bridport in Dorset. Although so young, Ferranti had already worked for Siemens, installing electrical equipment all over the country, had set up business on his own in a small factory in Finsbury to make dynamos and house-service meters, and had patented his first alternator. He was already well-known to all the leading figures in the electrical world, including Sir William Thomson, with whom Ferranti shared his alternator patent.

This was the formidable talent who took charge of the Grosvenor Gallery installation in 1886. In May Ferranti showed Alfred Bolton round the new generating plant underneath the gallery. Ferranti's large ideas soon led to the setting up of the London Electrical Supply Corporation Ltd in August 1887, with a capital of one million pounds. He envisaged an alternating current system which could virtually monopolise London's electricity supply, fed from a very large generating station. This could only be built outside the urban area of London, and Ferranti found a suitable site at Deptford. The customers would be supplied through the existing stations, which would become sub-stations, fed by a high voltage current from Deptford.

Work began in 1888 on this, the first modern power station. The gigantic proportions of the Deptford power station, with its 10,000hp engines and 10,000 volt alternators were ahead of their time. The design and execution of most of the electrical and civil engineering work at Deptford was Ferranti's own. On the question of the new 10,000 volt mains, he sought Alfred Bolton's advice. Ferranti came to Oakamoor on 3 March 1888 to discuss with him the best form for the leads to be used between Deptford and London. The mains were eventually made under contract and installed early in 1889. By the middle of that year, the smaller generating sets at Deptford, which could produce up to 5,000 volts, were ready.

Then came the first of the disasters which were to beset the project. The mains which had been supplied failed to carry even the 5,000 volts at which the supply started. This is perhaps not surprising, as this was well above the voltages previously used in public supply systems. Ferranti immediately set to work to design high voltage mains of his own. His design, which was entirely successful, was another major contribution to electrical engineering. The mains were to consist of two concentric copper tube conductors. The inner one was bound round with brown paper bands soaked in melted ozokerite wax to a thickness of ½in. This was then slipped inside the outer tube, which was drawn down onto the inner core, ensuring that there were no voids. The outer tube was earthed, and placed inside another iron tube. The whole of the mains, in rigid 20ft lengths, was put into wooden troughs filled with pitch. These were buried underground along the side of the London, Chatham and Dover Railway Co's tracks, and in underground railway tunnels. Forbes, the chairman of the London Electric Supply Corporation, and representatives of the railway company came to see Alfred Bolton on 18 April 1890, to discuss making Ferranti's mains at Oakamoor. Twenty-eight miles of the mains were made, and the supply from Deptford was restored, after a fire which destroyed the Grosvenor Gallery installation, on 15 February 1891.

Ferranti and the London Supply Corporation parted company in August 1891. But the Boltons maintained their links with him. He visited Oakamoor on 29 October 1891, and again on 23 February 1892. The connexions with electrical supply work meant that the firm was making more and more parts for dynamos and alternators, and electric mains. A large department was set up at Oakamoor making commutator bars for dynamos. These were drawn through dies, and again it was Thomas Bolton who was actively concerned with drawing the commutator copper from rolled rectangular sections. By the mid-1890s, Oakamoor already had over 1,000 different dies registered for this purpose. Large quantities of bare copper mains were also made, usually in 600 foot lengths.

Thomas Bolton and Sons were also involved in the next major step in power station development. By 1898, the Kensington and Knightsbridge Company of which Alfred was a director was finding that the load on its two generating stations was reaching 1,000kW. Extra plant was needed. They decided to have a joint venture with the neighbouring Notting Hill lighting company, and this led to the building of Wood Lane power station which started to operate in October 1900. This was notable for being the first three-phase generating and transmission system.

Electric Power and Traction

Electrification quickly spread from lighting to the railway and the tramcar. The first London electric tube services started in the 1890s, with the City and South London and the Central Line, using electric locomotives. The Bow-Upminster line followed in 1905, and Paddington-Westbourne Park in 1906.

But for Thomas Bolton and Sons it was the rapid growth of the electric tramway which was most important. Starting in Leeds in 1891, there was a wholesale conversion until by 1914 there were 264 electric tramway undertakings. The standard method of supplying current was through a trolley wire, the tram picking up the current through an arm on its roof. Other methods were tried, but this was the one which prevailed. Much of the trolley wire erected in the years before World War I was supplied by the firm.

Alfred Bolton had been approached in 1886 by Edward Blythe and asked to join the Electric Tramway Maintenance Company as a director. He was asked again in 1891. But although he did not actively take part in the electric traction business personally, he and his son Thomas made their mark. Trolley wire made new demands on the manufacturer of copper wire. The wire had to be of the highest conductivity, but at the same time great mechanical strength was needed to withstand the strains imposed by contact with the tram's connector arm. This could only be ensured by having as few joins as possible. Once again the ingenuity of the firm was sufficient to the task. In 1892, a new wire-bar mill was set up at Oakamoor which gave the firm a unique capacity to roll very long lengths of trolley wire, and gave them a lead in its manufacture which they maintained for a long time.

The engine of the new mill was started on 18 January 1892. High conductivity wire bars, billets, or 'bolts' as they were sometimes called, were received at Oakamoor, from Widnes or Froghall. They were made red-hot, and passed through a series of grooved rolls. By these a 4½in billet could be reduced to ¼in rod. The mill made a strong impression on the deputy editor of the *Electrical Review* who saw it working in 1897. He described the 'stumpy billet' becoming a 'squirming wire of great length' in scarcely a minute. At first the mill could only make rod from billets up to 135lb in weight, but this was soon to be increased. The firm guaranteed a variation in diameter of the rod of less than 0.0005in. Jointing was of course still necessary, but experiments in jointing overhead wire at Oakamoor had produced results that were absolutely reliable.

The Cable Makers

Thomas Bolton's were still making conductors for submarine cables. They made the wire for the West African Cable in 1885 and 1886. But with the growth of electrical supply, a new industry was growing up making power cables for overland use. Alfred was involved in the International Cable Co which did not prove very successful. But principally the firm supplied copper to the two new companies which began in the 1880s and 1890s. Callender's and British Insulated Wire.

Callender's Bitumen Telegraph & Waterproof Company Ltd was started in 1882 to exploit the founder's introduction of vulcanised bitumen for insulating cables and other purposes. The company started buying copper from Bolton's from the outset. The cable business began to flourish more rapidly after the 1888 Lighting Act and it increased more than four-fold by 1895. W.O. Callender visited Alfred Bolton for the first time on 29 October 1890, having just bought 20 tons of copper mains from him. In the next decade their association grew, and in 1900, Callender's new Cable and Construction Co bought 1,000 tons of copper from Thomas Bolton and Sons. Regular contracts for large amounts of copper followed. It was not surprising therefore that Tom Callender joined the board of Thomas Bolton and Sons in March 1902, just before the firm was reorganised and became a limited company.

The other major cable manufacturer to start at this time was the British Insulated Wire Company of Prescot. Its founder Atherton visited Alfred Bolton on 5 November 1891 shortly after the company started. Ferranti was also connected with the BIW Co, and it is probable that Boltons supplied them with copper although there is no direct evidence. Relations were certainly not as close as those with Callender. The Prescot firm in fact poached some of Bolton's wire-drawers from Oakamoor in 1900, with the intention of making their own wire. Originally nine men agreed to go to Prescot to start the wire-drawing shop there, and teach other workmen the trade. The terms offered were £3 a week for 54 hours, and an agreement not to join a union.

The growth of the cable companies was to have far-reaching consequences for the future of Thomas Bolton and Sons. But in the last years of the nineteenth century this was still only a small cloud on the horizon.

Froghall

The growth in demand for high conductivity copper was outstripping the capacity of the electrolytic plant at Widnes by 1890. To meet it an entirely new factory was planned, the first one built by the Boltons on a 'greenfield' site. It was decided to have it near Oakamoor rather than build an extension at Widnes and enquiries began in March for a piece of land to build on at Froghall, belonging to the Beech family. Made up of two meadow fields, it lay between the Caldon Canal to the north and the North Staffordshire Railway line and the River Churnet to the south. The site was to be served by a railway siding. Excavation began on 12 May 1890. The freehold of the site was bought by the partners in July for £600 17s 10d. Oakamoor at this time was said to have had a 'perpetual black canopy of smoke', so the Beech family had included in the agreement the stipulations that there should be: 'no smoke or vapour injurious to vegetation on any land comprised in the indenture..' and that 'no noxious refuse or other injurious substances shall be discharged into the River Churnet from the premises'.

By 21 January 1891, the depositing shed was nearly completed. Initially it was to contain two sets of tanks, arranged on different levels as at Widnes. Each contained eight tanks or vats, 5ft 3in long, 4ft 6in wide and 4ft 6in deep. By the end of February the two engines were nearly ready. Each engine was to drive two Elwell-Parker dynamos by a rope

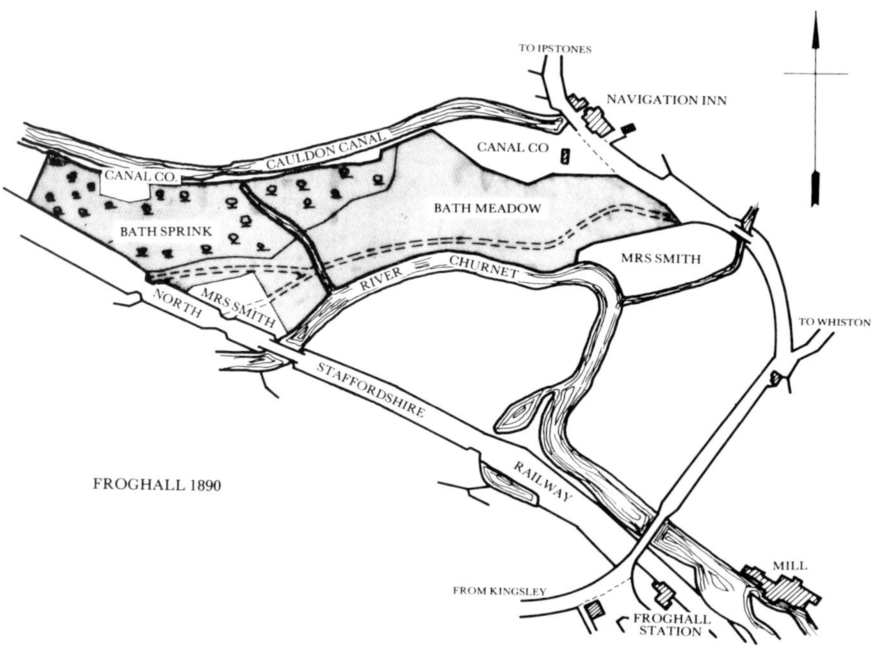

The land at Froghall before the works was built (based on a plan with the 1890 conveyance).

The steam engines at Froghall in 1897, showing the rope drives to the dynamos.

The Elwell Parker dynamos at Froghall for electro-deposition of copper. The engines were built on a higher level to give a longer rope drive.

drive. By early April the first trial depositing began, and on 11 April all four dynamos were running. Throughout April there were problems at Froghall. These did not come from the new electrical equipment, but from the engines supplied by Kirks, and Scott and Hudson. These triple-expansion engines were of the latest design, but the continuous use demanded by the electrolytic process caused difficulties with over-heating bearings.

The first electrolytic wire bar from Froghall was cast in the new refinery on 18 June 1891. It was taken to Oakamoor in the horse trap, drawn down and tested. It was found to have 98.1 per cent conductivity by the Matthiesen standard, a result which was thought good by Frank Bolton, the younger son of Alfred, considering that everything at Froghall was new, and that only two men brought from Widnes had any previous experience of electro-depositing. Apart from fish poisoning in the river, which caused the Beech family some anxiety, all was going well. It was immediately proposed to double the capacity, with two new engines and four more dynamos. Within a few years, the Froghall plant had a monthly refining capacity of 700 tons of copper, although a part of this was used for copper ore reduction, nickel-plating, electro-tinning of wire and other things.

The tanks were mounted in series rather than in parallel as experience showed that less copper was needed in the vats. In all there were sixty to seventy tanks, each one containing ten anodes and nine cathodes, totalling 180sq ft of depositing surface per tank. In their final form the tanks were smaller than the initial ones, being 3ft 6in deep, 4ft long and 2ft 6in wide. The conductor consisted of flat strip. After depositing was completed, the cathodes were melted down for casting into a suitable size. The billets were cast by hand ladling in open moulds. The mould was then tilted to allow the red-hot billet to slip out enough for it to be pulled out manually with a pair of long pincers. It was cooled in a stream of water and taken to Oakamoor.

The installation at Froghall was not cheap. The total cost of the works was £7,662. But the Froghall works became the main source of electrolytic copper for the firm in the years to come. The opening of new works led to a further extension of the partnership. In April 1892 Frank Bolton, who had been working at Oakamoor and learning the trade was made a full partner. Later in the year, Alfred discussed for the first time with his lawyer and banker, turning the firm into a private limited company. The question was discussed further in 1893 with the accountant Fred Whinney. But nothing came of these talks immediately.

Thomas Bolton and Sons and Aluminium

The importance of the firm's role in making copper for electrical purposes is already well known. Not so well known is the significant part played by Alfred Bolton in the industrial development of aluminium.

The history of aluminium is rather paradoxical. Although the commonest metal in the earth's crust, it was the last of the major non-ferrous metals to be mass-produced. This was because of the initial difficulty of extracting it from its ore. Deville first announced in 1854 a method of making reasonably pure metallic aluminium starting with alum, converting it to aluminium chloride, and then reducing this to metallic aluminium by the agency of sodium. This costly and protracted method was not improved on much until the early 1880s. By 1883, the total world production of aluminium did not exceed three tons, most of which was made in France. The price was an astronomic £5,000 a ton. Improvements in the chemical method of making aluminium were made by Webster and Castner, who formed the Aluminium Company Ltd at Oldbury in 1887, with a capacity of 45 tons of metal a year. In 1890 it was still the largest production plant in the world.

But the future of aluminium lay in the use of electrolytic methods for extracting the metal from its ore. This began in 1886 with the patents of Charles Hall and P. L. C. Héroult for making aluminium from alumina. These were followed in 1887 by the Bayer patent for extracting alumina from bauxite. The effect of these developments was to dramatically reduce the price to £1,792 a ton in 1887 and to just under £500 by 1889.

The first plant in Europe using electrolytic methods to make aluminium was set up at Milton near Stoke-on-Trent in 1887, by the Cowles Syndicate Aluminium Company. In this Alfred Bolton played a leading part. The story began in 1887 with a meeting at Oakamoor between Alfred and representatives of the parent American firm, Cowles Electric Smelting Co of Cleveland, Ohio, which was run by two brothers, Eugene and Alfred Cowles. The 'Cowles electro-thermal process' was discussed. This was a variation of Héroult's patent, which dissolved alumina in an electric arc furnace, using a carbon anode. This was better suited to making the alloys of aluminium, particularly the aluminium bronzes, in which Alfred Bolton was interested. These were known to have remarkable qualities of ductility and resistance to corrosion, as well as tensile strength. Alfred wished to use the alloys for wire and tube making at Oakamoor, for electrical and other industrial and military uses.

At the meeting Alfred was proposed as a director of the new Cowles Syndicate Co by P. Grahame Bourne of Brown and Boby. He accepted and the first meeting of the Cowles Co was held on 29 August. Initially, Alfred considered the site at Froghall for the works, which he went to look at on 2 September 1887. However, a position nearer the industrial area of the Potteries was preferred and work began at the Milton site by the Trent and Mersey Canal in November. The main building, 200ft long, 65ft wide and 20ft high, contained the engine, dynamos and sets of electric furnaces. The first shareholders meeting was held in the office at Milton on 29 December and by then the first set of twelve furnaces was finished.

By February 1888, the first aluminium had already been made, and Frank Bolton was trying to forge aluminium torpedo shells under the

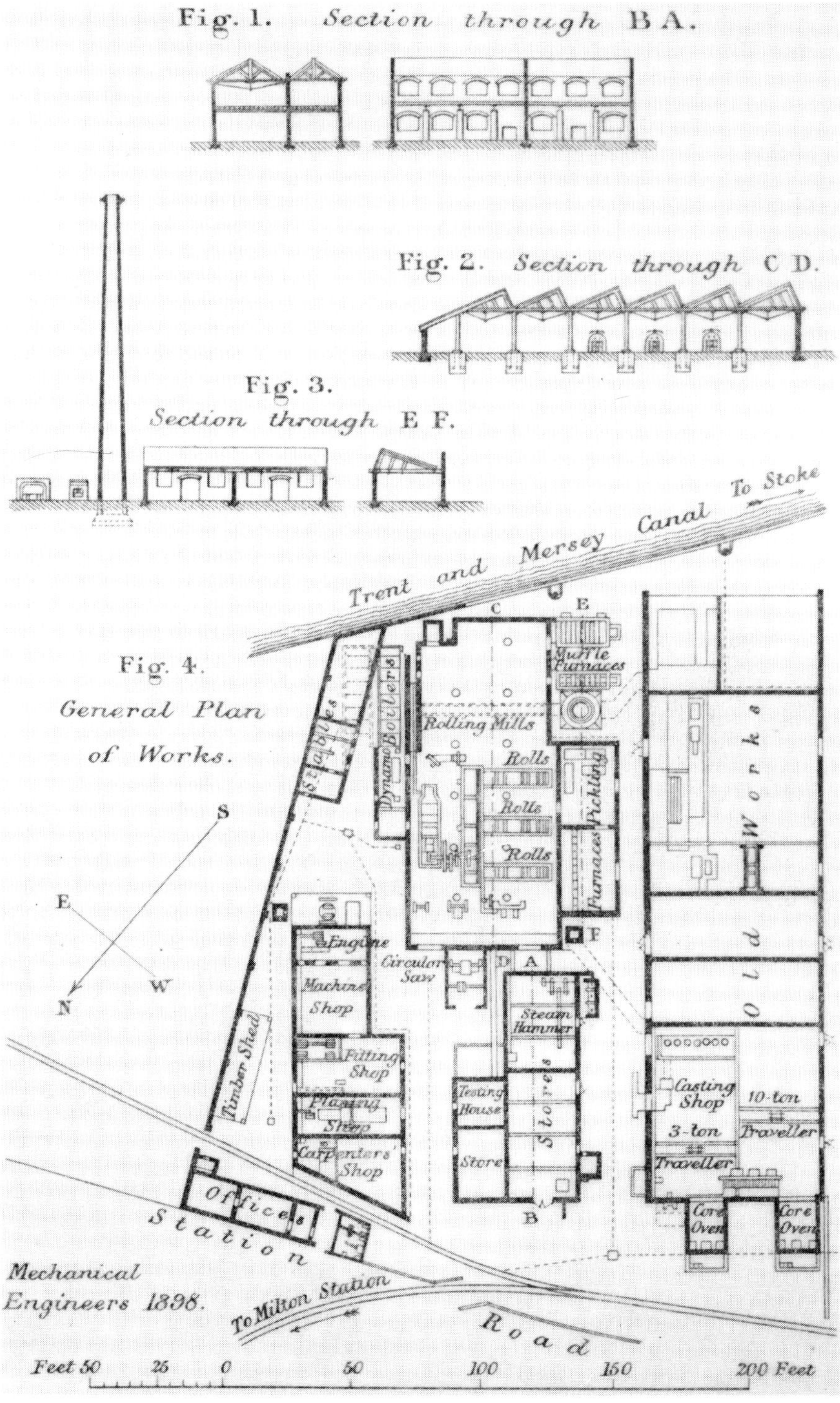

The Milton Aluminium works, near Stoke on Trent, 1898.

steam hammer at Oakamoor. Eugene Cowles himself came in March to discuss the making of the torpedo shells, and this resulted in an attempt to cast them in April. Later, in July, experimental work was done on aluminium bronze and aluminium brass. Cylinders of both alloys were cast at Oakamoor. Alfred attended the board meetings of the Cowles Co regularly. They were usually held in London, so he arranged them to coincide with meetings of the Kensington and Knightsbridge Lighting Co.

The American interest in the continuous wire-drawing machine, and the transatlantic origins of the aluminium process persuaded Alfred to visit North America. He landed at Montreal with Frank his younger son, on 7 September 1889. Their first industrial visit was to the Greening Wire Co at Hamilton, Ontario. This had been set up in the 1850s by Nathaniel Greening who had emigrated after retiring from his Warrington wire works. They had been among the first customers for the fine wire machine. This was followed by a visit on 23 September to the Michigan copper mines Calumet and Hecla. Here they saw the 5 per cent copper ore in granulated form being separated by flotation, and then refined. After visiting various wire mills, smelting works, and stamping works, they arrived on 4 October at the Cowles Electric Smelting Co's main plant at Cleveland. They met the whole Cowles family including the father, Edwin. He had known the Abolitionist John Brown, leader of the ill-fated attack on the Harper's Ferry arsenal, whose name was enshrined in the well-known Civil War battle song. Brown was also a native of North Ohio. The Cowles family themselves had helped to hide runaway slaves 'in the old days'. After seeing the Brush Electric Co works, the Boltons visited Irving Brown, the patentee of a weldless chain machine. This could make twenty-four links a minute from ordinary wire rod. This invention caught Alfred's eye, since no such device existed in Britain. On his return to England Alfred energetically tried to organise a syndicate of British wire-makers and others to exploit the Brown patent. Both men were impressed also by the Union Screw works, the principal manufacturers in America. There they found everything 'far ahead in modernness' of Nettlefolds or any English firm. Everything was done to save labour 'to a wonderful degree'. The Cowles other plant at Lockport was visited, and on the way back to New York, they called at several of the firms which had bought fine-wire machines, including the Coe Brass Co, the Aluminium Brass Company, and the Cabble Company of Brooklyn.

The following year the visit was returned by Alfred Cowles who came in May and studied the plant at Milton and Froghall. In 1891, Alfred for the first time attempted to make manganese bronze. He had visited Stones of Deptford, who were the principal British makers of this alloy in 1888. He now saw an opportunity to make use of the Milton furnaces for this purpose. Manganese bronze had many potential applications including locomotive tubes, locomotive axlebearings and other uses where its toughness, hardness and resistance to corrosion were valued. An upright furnace was prepared at Milton by Hobbs, who came over from the USA to make the modifications needed. Later in 1891, another expert from the American company came over to show the Boltons how to cast aluminium tubes. Five truck loads of these destined for the USA were shipped in August.

The Cowles Co continued to function throughout 1892, but by 1894 the works were closed down. Cheaper hydroelectricity made the steam-powered dynamos at Milton uneconomic. In May 1894, Alfred Bolton went to Milton to go through the books and put the accounts in order. The electrolytic manufacture of aluminium in Britain was now going to be undertaken by a new company, British Aluminium, which started up in 1894 with Alfred Bolton as a director. The Milton works became a refining and rolling mill for the new company, while the primary processing was done at Larne in Northern Ireland, Greenock and Foyers on the southern shore of Loch Ness, using electricity generated by harnessing the power of the Falls of Foyers. Aluminium was brought to Milton from Scotland for rolling into sheet. Alfred visited the new rolling mill after it had just started in December 1895. Unfortunately, he was not able to take a very active part in the new venture, due to his increasingly poor health.

The Robertson Piercing Machine

Naturally, electrical matters predominated in the firm's activities during this period. But Alfred was careful to keep in the forefront of the other aspects of the trade, especially tube-making.

The landmark here was the acquisition in 1892 of James Robertson's patents of 1884 and 1888 for making tube 'shells' in a hydraulic press. Robertson came from Manchester to see Alfred Bolton in July, and accepted his offer of £10,000 for his rights. He received £4,000 down and the rest over six months. This large amount of money gave the firm a machine which became an essential feature of tube-making at Oakamoor well into the next century. In November 1892, a new tube shop was

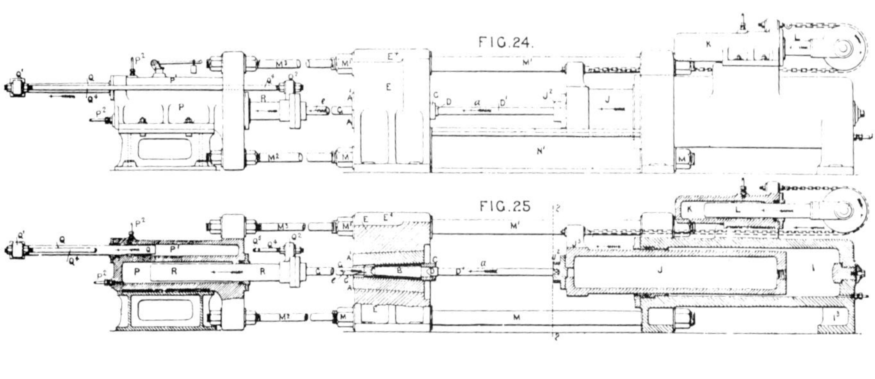

J. Robertson's tube shell piercing machine. From the 1888 patent drawings.

The Robertson piercing machine in the new tube shop at Oakamoor, built for it in The Meadow. This machine was the largest in Europe when installed.

The hydraulic pump at Oakamoor which powered three hydraulic draw benches and the tube piercing machine

started in Middle Meadow to house the monster, and the plant was ready in June 1894. Experiments were carried out, with Robertson present, and continued throughout the year until the machine was mastered.

To start the process, a billet of metal was placed in a tapered die, and was pierced by a mandrel driven into it with great force. The ram exerted a pressure of 2,400 tons, and the whole machine, which was laid out horizontally, had to be correspondingly massive to withstand such continual battering. The billet was softened or in a semi-liquid state, and the die was tapered so that the billet was not simply pushed through the mould by the force of the stem. Another part of Robertson's method was to revolve the die at high speed while the ram was working.

The machine was very flexible, and able to make cylinders from 3in in diameter, and with walls $^3/_8$in thick, to ones 12in in diameter and 1in thick. Locomotive tube 'shells' could be 'squirted' out from 8in billets in a few seconds ready for drawing.

The process made good quality tubes, and much cheaper than before. The purchase of the patent led to a big increase in Bolton's tube-making activity, involving brass and aluminium tubes as well as copper.

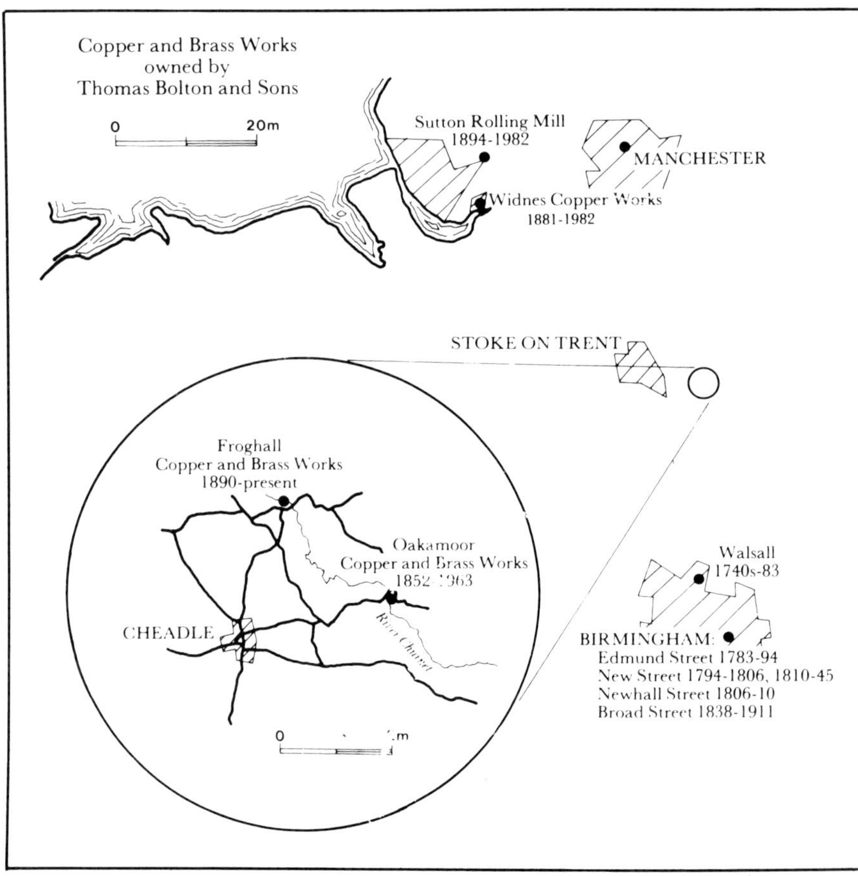

The Sutton works of Thomas Bolton and Sons, St Helens. The refinery is on the right.

Sutton

As the range of activities grew in Staffordshire, space was at a premium. One of the staple products was the calico printing roller, whose main market was in Lancashire. It made sense, therefore to move its manufacture to the textile area. In June 1894 Alfred and Thomas Bolton went to St Helens to see the Sutton works of Newton Keates, and watch their hydraulic roller plant at work. They were interested and went again to Liverpool to see about buying the works on 20 June. Alfred and Thomas finally concluded the purchase on 19 July.

Whether it was known to the Boltons or not, they were acquiring a distant relative. The founder of the works was William Keates, a former employee and protégé of the old Cheadle Brass Company. He left them in 1822 to become manager of the Whiston smelting works, and moved from there to manage the Spitty Copper works at Loughor in South Wales. As South American copper ore began to come into Liverpool, he revived the copper smelting industry in Lancashire, by setting up works in St Helens in the 1830s. The Sutton rolling mill was started in 1859 as part of his industrial complex which included works in the Greenfield Valley in Flintshire. The Boltons took over the business, which continued to trade as Keates and Bolton for some years. Much of the calico roller work was transferred there from Oakamoor.

Thomas Bolton and Sons by 1902

By the end of the century, the firm comprised five major production units: Broad Street in Birmingham, Oakamoor and Froghall in Staffordshire, Widnes and Sutton in Lancashire. Between them they produced a large array of products in non-ferrous metals. These can be summarised as follows:

OAKAMOOR

1 Largest output copper wire for electrical work (90 — 100 tons a week) from 2in diameter to $\frac{1}{100}$in.
High conductivity copper in tapes, rods, strips, and commutator bars.
Copper forgings for alternators etc.
2 Copper tubes, brazed and seamless from $\frac{1}{16}$in to 12in diameter.

The road past the works at Oakamoor at the end of the nineteenth century. On the right is the Oakamoor Post Office and beyond is the Coffee House Tavern. The latter is the only building shown here that survives.

Repairing a wall alongside the River Churnet at Oakamoor, during a period when the water was low, late last century.

A large output of finished and semi-finished brass tubes for boiler and condensers.
Aluminium tubes in considerable quantities from $\frac{1}{10}$in to 8in, as well as aluminium wire and rods.
3 Brass sheets, and wire, and solder.

WIDNES

1 Copper ingots and cakes of fire-refined copper.
2 Electrolytically refined copper for electrical purposes.
3 Copper sheets and plates up to 12ft wide, copper fire-box plates and copper rods for locomotive fire-box stay-bolts.
4 Large copper pans made by the traditional battery methods, especially for mining purposes.

FROGHALL

1 Electrolytically refined copper for Oakamoor.
2 'German silver' or nickel-plated copper made by electro-deposition.

3 Electro-tinned copper wire and other products.

BIRMINGHAM

1 Chiefly brass-wire and rolled brass.

2 Specialities: fine wire of all kinds especially for weaving wire for paper mills.

ST HELENS

1 Calico-printing rollers and other copper cylinders.

2 Rolling copper sheets and plates.

The Broad Street works were now comparatively small, with about 150 workers. The Sutton mill employed about 300, and Oakamoor about 450, with another 50 at Froghall. Widnes had the largest workforce with 600 to 700 men, since the work there was heavily labour-intensive. All the casting was done by hand, and so was all the manipulation of sheets and plates in the rolling mills.

By 1901, Alfred Bolton had become a patriarchal figure in the local community. His political interests as a County Councillor had taken up more and more of his time of late, and he served on numerous committees. Following his interest in scientific education, he helped to found the North Staffordshire Technical College in Stoke-on-Trent, buying the land for it with his own money.

But he was now 73 and in ill-health. He returned from a visit to London on 21 November with a bad cold, and was poorly for the rest of the month. He managed a day's shooting on 27 November, but immediately became worse, and died at 8.30am on 2 December 1901.

His death produced an immediate financial crisis in the firm. On 3 December Charles Toller went to London to deal with 'business worries with the Bank'. The day after Alfred Bolton's funeral, there was a 'very stormy and troublesome meeting lasting till near midnight', between some of the partners and the bank, and a list was made up of the property and shares belonging to each partner. The firm was put into the hands of a receiver, Whinney, 'pending further arrangements'.

Part of the problem was that personal and business property were hopelessly entangled, because Thomas Bolton and Sons was a private partnership. Until Alfred's will was proved, the exact extent and distribution of the assets could not be easily determined. These problems were compounded by other more personal developments. Thomas Bolton, the heir apparent of the firm, had left Oakamoor and his wife Nina in August 1901, and went to America with his new love, Connie Brown. A young governess, said to have been rather ugly and of 'masculine appearance', Connie nevertheless shared Thomas's taste in music, which had always been an abiding passion in his life. After a visit by a friend, Thomas returned to Oakamoor, but left again because of the friction in the family caused by his desertion of Nina. Thomas was thus not present when his father died, nor during the flurry of meetings which took place afterwards.

A loan of £275,000 from the Capital and Counties Bank was arranged to meet the creditors' demands. Early in 1902, the decision was taken to transform the partnership into a limited company, and bring in directors from outside the family circle. How far this was an enforced decision cannot now be determined, but it made obvious sense in the circumstances.

Alfred Bolton's estate was valued at £248,321 12s 10d after his death. The increase in the family fortunes since his father's day is a fair measure of the success of the family firm. The farms and other holdings around Oakamoor and elsewhere were divided mainly between his two sons Thomas and Frank. Alfred's share of the business was to go to both brothers in common, as did all his other shares and investments. But the main legacy was the standing of Thomas Bolton and Sons as one of the world's leading firms in the manufacture of copper for electrical purposes.

CHAPTER FIVE

Some Twentieth-Century Landmarks 1902-83

The death of Queen Victoria marked the end of the era of industrial leadership for British industry. A new and bleaker age began with increasing business difficulties brought about by competition from Europe and North America. Sharp changes in market conditions from boom to slump became characteristic. Normal business activity was massively interrupted by two world wars and their aftermath. Rumours of wars led to a greater involvement with military contracts.

All this was felt by Thomas Bolton and Sons Ltd in the years before World War I. The profit figures for those years show a rapid rise to a peak of £89,758 for 1907, and a sudden collapse to £25,604 in 1908, followed by another climb to £70,011 in 1912, and another collapse. During these years, British cable makers were forced out of European markets by the

The Oakamoor works nestling in the Churnet Valley, probably about 1910.

Maintenance work on one of the steam engines at Oakamoor about 1912.

growth of tariff protection. This had a serious effect on Thomas Bolton and Sons Ltd, because of their increasing involvement with Callender's company. One of the innovations introduced in the new limited company was the presence of T.O. Callender on the board. His father had visited Alfred Bolton in 1891, as was mentioned in the last chapter. From the reminiscences of an Oakamoor man who as a young boy carried Callender's bag to the station on that occasion, it appears that he had come to discuss with Alfred the merging of their two companies. This was not to happen, but there was a 'gentleman's agreement' between them, which governed their relations more or less up till the merger of the companies after World War II. The agreement was that Callender's guaranteed to buy their copper requirements from Bolton's, on condition that Bolton's did not attempt to enter the cable-making business. In the long run, keeping to this agreement probably worked against the interests of the senior and longer established firm. But at the time it arose naturally from the different characters of the two companies. Callender's strong point was salesmanship, which became a priority as they were forced out of the European markets, and had to seek new customers all over the world. The Boltons' forte lay more on the technical side. The family were reluctant to spend money on advertising and agencies until after World War I, when the post-war recession forced a change in attitudes. The concentration on technical aspects of the business remained the main characteristic of Thomas Bolton and Sons Ltd. They maintained a very wide range of products, and liked to specialise in meeting any particular requirements their customers had, rather than selling a relatively standardised article like cable.

The Bolton family continued to dominate the firm after it became a limited company in 1902. It was still a private company, and its ordinary shares were never quoted on the Stock Exchange. Thomas Bolton returned as local manager at Oakamoor for a while, but finally decided to move away with his new wife Constance to Tunbridge Wells in 1905. He was thus removed from day-to-day control of the company's main works in Staffordshire. He worked primarily in the London office, and was finally elected Chairman in 1909, after the resignation of Sir George Robertson, the outside chairman who has been brought in in 1902. He became Joint Managing Director in 1919, and remained the leading personality in the firm until he lost his executive powers in October 1931.

Thomas Bolton's three sons also began careers in the company in these years. Michael, the eldest, became manager at Oakamoor in 1905, and in 1911 was made 'local director', a post peculiar to the firm, at Oakamoor and Froghall. Edward was sent to Birmingham in 1908, and then moved to the Froghall Brass Works in 1911 as local director when the Birmingham works was closed and its functions transferred to Staffordshire. In 1924, he went to Widnes, where he remained until his early death in 1939. David Bolton started as the engineer in the Froghall Brass Works in 1911. After World War I, he took charge of the company's diamond die works in Dorset, of which more later.

The new board had to make changes in the traditional Bolton methods of financial control. In the past each of the works had been largely autonomous as far as accounts were concerned. But now the board set up a Costs Department to give them an independent view of the financial side of operations. There were objections to this, not least from Thomas Bolton himself. He claimed his figures for Oakamoor were more reliable than those of the Costs Department, but acquiesced in the new watchdog as it 'kept everyone on their toes'. It is remarkable that many large Victorian companies had no very clear idea of their costs. While they enjoyed an unchallenged dominance in the market, there was no great incentive for them to bother about them. But as the cold winds of competition began to blow harder, costs could no longer be ignored.

Another sign of changing times was the closure of the Boltons' original factory in Birmingham. This was decided on in September 1910, and the works were gradually run down until the final closure in early 1912. In the following year, the buildings were demolished, and the site sold to Guest, Keen and Nettlefolds in 1916. By the early years of this century, there was little point in having a separate works with products basically the same as those at the Staffordshire factories. The product line, mainly brass wire and sheet, was removed to Froghall, where a new department was prepared, known as the 'Brass Works'. This came to include also work with other copper alloys apart from brass in the 1920s.

Wire continued to be the main product from Oakamoor. Post Office

Thomas Bolton, 1858-1937.

telephone and telegraph wire was kept in the store there until it was needed and then shipped from Oakamoor around the country. The firm was able to win a contract from the German Post Office for 300 tons of wire in 1907, although this was the last from this institution. Another 400 tons was made for the Japanese State Railways.

The manufacture of trolley wire improved further in 1909 with the building of a new breaking down mill at Oakamoor. The earlier mill, started in 1893, had led the way in making long lengths of thicker wire

Building the brass works at Froghall just before World War I.

The Froghall fitting shop 1915. This department was moved from Birmingham in 1912.

2,000hp Belliss and Morcom triple-expansion steam engine Nora, *named after Miss Nora Bolton. This was installed at Oakamoor in 1917 to drive 500V DC generators.*

The breaking-down mill at Oakamoor. This was built in 1903, and this photograph shows it after 1910-11, when it was altered to take the large wire bars of one ton (seen here), used to produce continuous trolley wire.

possible by rolling wire rods directly from wire bars. At first this was limited to high conductivity wire bars weighing less than two hundredweight. Thomas Bolton saw the advantage of being able to roll from larger wire bars, and the new mill was built especially for this. The first rods were rolled in January 1909 from a wire bar weighing 7½cwt and by the end of the month, this had risen to 11½cwt. These high conductivity wire bars were not of course commercially available, and were made at Widnes or Froghall. By 1913 2270lb could be rolled and drawn into one length, and it was possible to make a mile of trolley wire without joins. This mill was modernised and remained in use up to World War II. The original rod-rolling mills of 1887 and 1893 were finally replaced by the new combined rolling mill at Froghall by that time. Many contracts were undertaken for major authorities who were setting up tramway systems. These included the London County Council, Manchester, Liverpool and Brighton, as well as more exotic demands from Egypt.

Another new departure at Oakamoor was the drop forging shop started around 1903. Here were made the components for motors and other electrical equipment which were in ever-increasing demand. This led to an expansion of the machine shop at Oakamoor where the parts could be finished. The making of machined parts from high conductivity copper and copper alloys gave Thomas Bolton and Sons Ltd an important new production facility.

Bronze trolley wire.

Reels of trolley wire.

Tube-making was improved by the introduction under licence of the German Heddenheimer process. The company was very concerned to improve its methods for manufacturing artillery shell bands, for which they had had regular contracts from the Army and the Navy.

The Mersey Copper Works at Widnes became the official head office of the company after 1902. The main activities of smelting and refining, and of manufacturing copper into plates, sheets and other forms continued. The majority of Bolton's copper was purchased from the American producers. The activities of the notorious Copper Producers Association in the USA made the life of the copper buyer Charles Bolton Toller very difficult. The price was kept at an artificially high level by the Association until 1910, when a sudden temporary collapse of the ring tempted the company into large borrowings from their bankers, Capital and Counties, in order to finance considerable over-buying of copper while it was still cheap. Anxious to break the strangle-hold of the Americans on the copper supply, the board began to look towards the newer Central African sources, and expressed interest in 1918 in taking over the Katanga Contract. However, a shift to African copper did not really begin until the late 1920s.

The rolling mills at Widnes continued to be among the main suppliers of locomotive copper firebox plates and other items to the railway companies, with regular sales to the Midland, London and South Western, the Great Western, the Caledonian, the Great Central and the Great Northern.

Large diameter special purpose copper tubes at 'The Meadow', Oakamoor.

An entirely new venture was begun at Widnes in 1903, which broadened even more the range of products offered by Thomas Bolton and Sons Ltd. This was the manufacture of copper sulphate. In the 1870s the vineyards of Southern France were struck by phylloxea, a disease which threatened to destroy France's most prestigious product, and the rural economy of the country. By the 1880s, wine output fell by 75 per cent. The application of copper sulphate in solution was found to protect the vines against the disease, and this led to a big demand for the 'Bordeaux mixture' in all the wine-growing areas of the Mediterranean. Copper sulphate was found to have a more general application in defending crops against various fungus diseases, and so a substantial world market grew up.

The copper used in the sulphate works was reclaimed from residues, both in the Mersey Copper Works and others in the area. It was granulated into copper shot and then dissolved in sulphuric acid, readily available from the local chemical industry. The solution was then cooled in lead vats, and the copper sulphate crystallised onto lead straps suspended in the liquor. It was a useful product for Boltons as it used up copper that would otherwise be wasted.

Another new development before World War I was the manufacture by the company of its own diamond dies. These played an important part in building up the reputation for quality which Thomas Bolton and Sons Ltd enjoyed. Diamond dies for wire-drawing in the company's works were in constant use for up to seven years without the need for resetting, due to the hardness of the stones. Up till 1906 the company had bought in their diamond dies, which were worth many thousands of pounds. In September 1906, it was decided to purchase the small works of one of their British suppliers, W. Neder, at 208 Birchfield Road, Handsworth, in Birmingham. In fact the works was only leased until 1917, but the company now became the only British manufacturers of diamond dies, which they supplied to others as well as to their own works. It was of course the hardness of diamond which made it useful for wire-drawing dies, before the development of tungsten and other alloys in the 1920s. Only diamond could cut diamond so it was essential to use diamond powder to bore the holes. Small steel drills were used, fed with oil and diamond powder, and made to turn for many hours. After being pierced, the holes were shaped to suit the kind of metal being drawn. They were then polished to the required gauge, and adjusted to within one ten-thousandth of an inch. The diamond was then mounted in a brass case a little larger than 1in diameter, in readiness for use.

World War I meant that Thomas Bolton and Sons Ltd went over to munitions production for the duration. The markets for some products like copper sulphate were cut back as a result of the war, and as the years passed the Government stepped in to restrict the commercial production of copper wire in order to preserve vital stocks of the metal. One of the main specialities of the company was the manufacture of copper shell bands, and output of these reached half a million a week at its peak in the

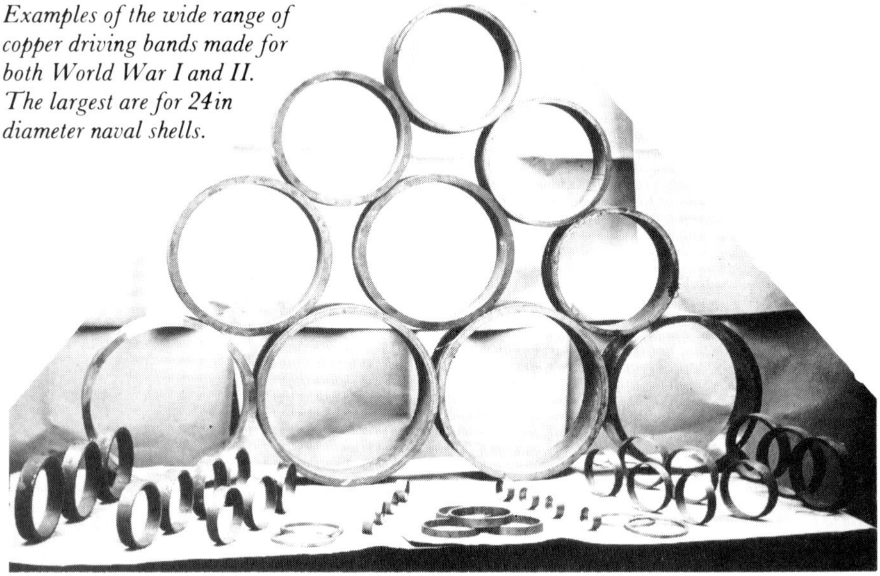

Examples of the wide range of copper driving bands made for both World War I and II. The largest are for 24in diameter naval shells.

war. The Oakamoor plant could cope with bands for shells up to the 24in needed for big naval guns, and was one of the few in the country capable of producing them. Cartridge cases of all kinds were also made. One casualty of the war was the electrolytic plant at Froghall. In order to make room for war-time production, the depositing plant was demolished. The lead from the vats was realised, and the nickel plant was removed to Widnes. Extra strip mills were to be put on the site when needed.

As was the case with many war industries, the firm made considerable profits during the conflict. The Government allowed 65 per cent write-offs against plant, like the 600 ton Fielding and Platt horizontal extrusion press installed in 1916. In the event, large reserves set aside to meet expected excess profits duty was not needed, and £200,000 was distributed in shares to increase the capital value of the firm. Most of these went to the existing shareholders.

Like much of British industry, Thomas Bolton and Sons Ltd enjoyed a brief post-war boom as a result of the destruction of European competition. During this time more investment was made at Froghall in the form of a new wire mill costing £6,000. From now on new equipment in the Staffordshire factories was to be largely concentrated at the Froghall works.

In 1922, profits fell to their lowest ever levels since 1902, at £22,901. This compared with £120,808 for 1920. Part of this was due to losses during the nine-month coal strike. The following two years showed net manufacturing profits of only £54,906 and £43,747, as French and German industry began to get back into full production. To meet this situation, a new stress was laid on sales organisation. Thomas Bolton took

The Bronze wire mill ('Yankee mill') at Oakamoor, built on the site of Patten's gas works. This photograph shows it after a disastrous fire in 1921. The three-storey building was later rebuilt as two storeys, and used for continuous casting.

charge of home and export sales departments at the company's London office.

In addition to his duties as company chairman and joint managing director, Thomas Bolton now became the first chairman of the newly-

Copper bars 3-5¼in diameter with the rod mill in the background.

formed British Non-Ferrous Metals Association. This body was primarily concerned with promoting the interests of non-ferrous metal producers and manufacturers, and co-ordinating research. Thomas Bolton continued in this post until his death in 1937. In the 'twenties the company began to expand its own research facilities. The leading figure in this was W.E. Alkins, later Dr Alkins. The son of an Oakamoor foreman, Dr Alkins was educated at Leek Grammar School. The company decided to contribute a loan of £25 to enable him to take a course at Manchester University. He thus became Oakamoor's first graduate. The old Cheadle Brass company had helped William Keates in the same way just one hundred years before. But in Dr Alkins's case, the investment was returned a hundred-fold. He returned to the village to work for Thomas Bolton and Sons Ltd, and in 1928 was put in charge of the new research laboratory then being built at Oakamoor. One of the main concerns of the company and the Association was the new range of copper alloys being introduced in the years after the war. These were alloys which involved small additions of other elements to copper. As Thomas Bolton himself pointed out in a speech to the Non-Ferrous Metals Association in 1932, there was no basic knowledge which enabled metallurgists to predict the performance of an alloy in the same way as they could that of a mechanical structure like a bridge. There had to be more fundamental research into these matters. Not all of these alloys were for electrical purposes. Copper with 0.5 per cent of arsenic in it was being used more and more for firebox plates and other purposes. Arsenical copper was first discovered by Park, the observant chief engineer of the North London Railway in the 1880s. He found that the fireboxes which lasted the longest proved on analysis to contain traces of arsenic. This later became standardised at 0.5 per cent.

Most of the new alloys however were concerned with giving copper qualities useful to the electrical engineer. Thomas Bolton and Sons Ltd pioneered the use of many of these, particularly cadmium copper. Addition of this relatively rare metal greatly enhanced the tensile strength of copper. The introduction of this alloy enabled the company to make the conductors for the Thames Crossing, a major engineering operation to take the new National Grid cables across the Thames. Unfortunately for Thomas Bolton and Sons Ltd, the 132,000 volt system generally used aluminium conductors with steel cores. The contract won by Callender's in 1928 for the South-Eastern section of the new grid did not therefore lead to a bonanza for their copper suppliers. However, Thomas Boltons were the only firm in the country capable of making conductors in the unbroken lengths required for the crossing at Dagenham. The span between the supporting towers was 3,060ft, so the seven cables had to combine great strength with the capacity to carry the 132,000 volt current with complete reliability. The solution arrived at was to have seven tin-bronze wires in the centre, surrounded by eighty-four cadmium copper wires of the same size. Each of the wires was made in one 6,000ft length without joints in the unique mill at Oakamoor. Another contribution to the new electrical supply system made by Thomas Bolton and Sons Ltd was the development of cellular conductors for transformers. These were in fact stranded copper tubes, which

The diamond die works at Parkstone, Bournemouth.

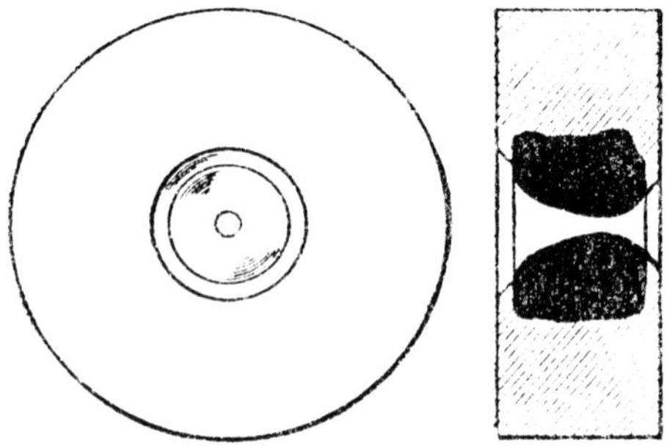

One of Thomas Bolton's diamond dies made at Parkstone.

combined lightness with the strength required, and were patented by the company.

In 1920 the diamond die business was moved from Birmingham to Dorset. David Bolton had left the Froghall Brass works and gone south to live with his mother Nina Bolton in Bournemouth. There he found a suitable site in nearby Parkstone, and he found a new role in the company as local director of the new works. Thomas Bolton and Sons Ltd were still the only British manufacturers in this field, although a considerable number of diamond dies were imported into Britain. These dies were now widely used for wire-drawing down to a diameter of 0.0004in, the smallest dies being used for drawing tungsten filaments for electrical lamps. All types of wire were now being drawn through diamond dies, from telephone and telegraph wire, to weaving wire for cloths, brush wire, pin wire, surgical wire, and gold and silver thread for braid.

From a peak in 1925 of £103,469, manufacturing profits fell to £35,937 in 1927, a figure which included £15,000 transferred from the Reserve Fund. However, in the three years which followed the fortunes of the firm were lifted by the boom in the United States. Stock and commodity prices rose steadily, and this included the price of copper. Boltons benefited in two ways. Firstly, they 'over-bought' very heavily, that is they bought more copper than they needed for their own purposes. This was done so that the copper could be resold later at a higher price. Secondly, they were able to increase their profits on the sales of copper which they refined electrolytically at Widnes. The price of electro-copper went up much faster than standard copper, and so it paid to buy more standard copper and refine it for resale. As the price rose above £100 a ton the returns from these activities were considerable, and were reflected in improved profits.

The effect of the collapse in commodity prices in 1929 was therefore very serious for the firm, both in the short and long term. The fall in copper prices began in mid-1929 and the profits of the earlier part of the year turned to heavy losses by June. The crash was particularly traumatic for Charles Toller, the company's copper buyer for forty years. The over-bought position of the company he regarded as his responsibility, and tragically took his own life. Losses however continued as the copper price plummeted to £40 a ton. The company was now left with large stocks of copper bought at much higher prices which they had to sell at a loss. In October 1929 came the Wall Street crash and the onset of world recession which cut sales and left the firm very vulnerable.

There had already been substantial rationalisation in the British copper industry. Many of the seven firms in the Manufactured Copper Association after World War I had disappeared into the recently-formed giant Imperial Chemical Industries. As the largest independent producer of high-conductivity and manufactured copper, Thomas Bolton and Sons Ltd was an obvious target for take-over in its present weakened state. The first offer came from abroad. In 1930, the major French concern Trefileries et Laminoirs du Havre (the Havre Wire and Rolling Mills), offered to buy a controlling interest at 11 shillings a share. At first the directors rejected the bid, but the company secretary Matthews persuaded them to reconsider. The threat of a takeover then encouraged a counter-bid from Callender's Cables and British Insulated Cables, the successor to the Prescot British Insulated Wire firm. They offered to buy 100,000 of the £1 ordinary shares at 12 shillings each, and 160,000 at 13 shillings, plus an option to buy another 60,000. The offer was accepted by the majority shareholders Thomas Bolton and his brother Frank. This was the beginning of the end of the Bolton family's control of the firm. The interest of the two companies was natural enough. Callender's Cable and Contruction Company took 50 per cent of Thomas Bolton's output of high conductivity copper in 1930. BIC also had a similar product range to Boltons. After the share purchase an agreement was reached between them over the marketing of copper rail bonds which both companies were selling. As owners of the controlling interest, Callender's and BIC nominated their representatives to the board. Sir T.O. Callender continued as director, with Charles Pipkin as his alternate. British Insulated Cables nominated Brigadier-General Critchley to look after their interests, along with John Ferguson, their Deputy Chairman, and G.H. Nisbett, their Managing Director. BIC, like Boltons, refined their own copper, and rolled and drew wire for the cables which they manufactured.

The final reorganisation took place in October 1931. Thomas Bolton became a non-executive director, but continued as chairman of the board. George Rathbone likewise lost his managerial functions but continued on the board. Frank Bolton remained a director at the wish of the Callender and BIC representatives. Hugh Bolton Toller, who had taken C.B. Toller's place at Widnes became general manager over all the works. BIC brought in one of their men, C.F. Davey, to run Widnes, while Michael Bolton continued as works manager in charge of the Staffordshire factories. Edward Bolton kept his responsibility for sales. The youngest of Thomas's sons, David, had intended to resign from the company at this point, but in fact continued for some years in charge of the die works.

The new regime intended to make changes to meet the unprecedented situation facing the firm and the industry. An 'advisory executive' was set up, which included Nisbett and Brigadier-General Critchley, to make recommendations. One of the changes which they made was to have regular reports from each of the executive directors at every board meeting. Previously a more gentlemanly atmosphere prevailed, where no-one was obliged to speak by the form of the agenda. In addition the board insisted on authorisation of all capital expenditure, however small, in order to keep down costs. G.H. Nisbett, in particular made detailed comparisons of productivity as between the Staffordshire factories and his own works at Prescot, pointing out that Oakamoor's wage costs were £14 a ton produced, whereas at Prescot a ton cost only £5.27 in wages. It was pointed out in reply that Bolton's factories made a very wide range of products, some made in small quantities to specific customer requirements, like commutator segments. These needed more labour-time than large-scale wire drawing of fairly uniform gauges, which was the practice at Prescot. There was of course truth in this, but in the next three years an improvement in the productivity of the Staffordshire factories was achieved, largely by shedding labour.

There were inevitable casualties in the slump. One was the closure of the electrolytic depositing plant at Widnes. Electro-refining of copper had now become uneconomic, due to a combination of factors. 'Argentiferous' copper was now being refined in the countries of origin, so there was no longer any profit to be gained from the extraction of silver and other precious metals. The price of electro-copper had fallen to such a

The Oakamoor cricket team, probably with the Sentinel *cup which they won three times in succession in the 1930s. In the centre holding the cup and wearing a blazer is Michael Bolton.*

Francis Bolton (1866-1951) and his wife at their home at Moor Court, near Oakamoor.

level that in any case, this relatively expensive process would have probably been stopped. Most of the facilities at Widnes were removed in 1932. Another blow to Widnes was the temporary closure of the sulphate plant, also hit by the consequences of the agricultural slump world-wide. However, it proved possible to re-open half the works in 1931 on a seasonal basis only.

A near-victim of the times was the Parkstone die works. This was threatened by the influx of European dies into Britain. The board decided however to give a few months' reprieve to see what the effect of impending tariffs might be on sales. The reprieve was extended as trade began to improve. Eventually, by January 1933, the company was persuaded of the viability of the department, and it was transferred from

Parkstone to Froghall, still under the management of David Bolton, who had withdrawn his original intention of resigning.

In order to save money, the board considered the closure of the Sutton works, and the removal of its activities to Widnes. On looking into the costs involved, however, the board found that they would have to spend £18,000. The proposal was deferred.

As well as retrenchment, including wage-cuts, the company sought out new opportunities. A new electric furnace was bought for Froghall to make the oxygen-free copper which was coming into use. Another venture was the manufacture for the first time of small diameter tubes for domestic use. During the slump there was a paradoxical growth in house building as real incomes rose, and this gave rise to an increased demand for building materials. Other firms, including the Broughton Copper Company, were already making copper tubes for the domestic market, and a scheme for joint production with them was proposed in 1930. Talks proceeded throughout 1931, but by December they had broken down, and although the idea was revived at different times throughout the 1930s, nothing came of it. Small tube production nevertheless went ahead, and new long drawbenches were installed at Froghall. In 1932, another £6,000 was authorised for a new extrusion press at Oakamoor for domestic copper tubes. Boltons were amongst the first to use this kind of equipment for small tubes. The production was expanded in 1935 to get a larger share of the 18,000 tons per year market. It was always recognised, however, that the company could only achieve a marginal share.

Another new development took place at Sutton. In 1931 it was agreed by the board to stop production of copper engraving sheets, and to consider making instead the newer zinc sheets, to be cast at Widnes and rolled at Sutton. BIC agreed to give Thomas Bolton and Sons Ltd all the information they had on these. Investigation showed that all zinc sheets for the printing industry were being imported at that time, so it was a promising market. By March 1932, an investment of £14,000 was authorised to begin manufacture of zinc sheets, at the rate of 3,000 sheets a week, using two shifts. After some preparatory trials and negotiations with the printing trade association, the zinc sheets were accepted as equal to the North American product. The marketing arrangements were later handled for Boltons by SD Syndicate in an agreement signed in 1938.

In 1930 it was noticed that the use of synthetic dies for wire-drawing was on the increase at the expense of the older diamond dies. David Bolton at Parkstone was asked to look at the new alloy dies. By the time the works was transferred to Froghall, 'Ardoloy' dies were being made alongside the established diamond. The tungsten carbide alloy used was a patent of British Thomson Houston, and the dies were made under licence from them.

The next major proposal for reorganising the works was the closure of Oakamoor and the removal of its machinery to Froghall. This was first suggested in 1934, and was given very serious consideration. Detailed plans and costings were made. The main argument was that the activities of the two works largely overlapped, and the expense of two establishments was a luxury the firm could not afford. Most of the post-war investment had been made at Froghall, and there was room to expand there, whereas the Oakamoor site was now over-crowded. New machinery had been bought for Oakamoor, for example a new bolt mill in 1933, and some new wire-drawing machines. But the company was concerned about the 'dilapidated' appearance of the Oakamoor works, some of whose buildings dated back to the eighteenth century.

As trade began to pick up again after the slump, ambitious rebuilding schemes seemed justified. While the plan to close Oakamoor did not bear fruit at this time, extensive new investments were planned for Froghall. The first of these was the new casting shop and its related stamping and machine shop, which were begun in 1934, and were in operation by the following year.

The new casting shop was needed to cope with the wide range of alloys now in demand from the firm's customers. Cadmium copper was in increasing demand, and was thought likely to replace copper generally for line wires. Another growing interest was in the mechanical qualities of copper alloys for use in internal combustion engines. Phosphor bronze was used by the Bristol Aeroplane Company for valve guides in their aero-engines, and aluminium nickel-bronze in the valve seats of oil engines. Aircraft undercarriage components were made from manganese bronze. In addition gun-metal, silicon bronze, cupro-nickel and nickel silver were cast in the new shop. Cadmium copper was cast in ingots up to 14½cwt for drawing into trolley wire.

The stamping and machine shops were located in opposite ends of the same building. The drop-stamping plant was able to produce a variety of specialised items. Most of the building was taken up by the machine shop, where material from the drop-forges or stock from other departments in the works was worked on. The new machine shop could make components for a great variety of purposes, in the typical Bolton style. Aircraft parts were forming a larger and share of the output of the shop.

This investment enabled the firm to attract important contracts from the Government as re-armament began to get under way in 1935 and 1936. The growth in business as a result of defence work encouraged the board to continue expansion. The next major step was the combined rolling mill, on which work began in 1936. This plant, expected to cost £100,000 in buildings and equipment, was given a £50,000 grant by the Government. The new rolling mill was designed to replace many of the existing departments both at Froghall and especially at Oakamoor. In September, the first component of the mill was ordered from Schloemann's of Dusseldorf for £19,000. This was the most modern equipment of its kind then available. The building and the combined mill was finished and working in early 1937.

The scene in the valley was now very different from what it had been only a few years before. In 1934, there were 800 workers, doing a day shift only. By mid-1936, there were already 1,400, and by the following year

The Oakamoor works from the air, 1949.

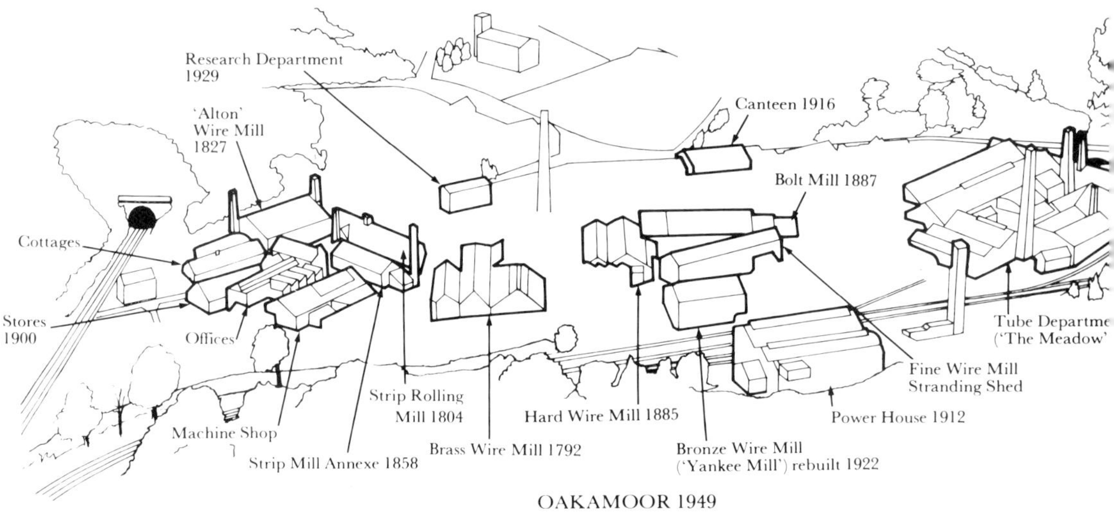

Key to aerial photograph, identifying some of the major buildings.

Multi-hole wire-drawing machines, in the bronze wire-drawing mill at Froghall.

the works were going day and night in three shifts largely due to defence work. Aeroplane engine valve-seatings, and copper shell bands were among the main items produced.

Elsewhere, the situation was not quite so happy. The commercial products of the company were all subject to the quotas laid down by the various trade associations. In manufactured copper Boltons were allocated 12 per cent of the market. They had a much larger share of the High Conductivity Copper Association's total, naturally. But commercial markets were still below their pre-slump levels. Rationalisation of Boltons' large-tube and shell output was considered in 1935, and the board were prepared to offer it to ICI if they made an acceptable offer. Similarly in 1937 the sale to ICI of the whole Manufactured Copper Association's quota, including the locomotive work and the sulphate plant at Widnes was proposed. However, the chemical giant made no offer on either score. Very little investment had been undertaken at Widnes in the recent past. The timber staging which supported the cooling vats in the sulphate works, for example, had not been renewed for thirty-one years and by 1937 was in danger of imminent collapse. Expenditure of £33,000 was authorised in 1938 to make improvements at Widnes, including a machining shop for cake copper and an extension to the rolling mill. The casting processes however remained unchanged. Hand-ladling in the refinery shop continued much as it had done since the nineteenth century. It proved difficult to mechanise these labour-

a

Three photographs showing the growth of the Froghall works:

a *1902 with the electro-deposition sheds in the centre.*

b *1912 showing the brass works and chimney in the centre.*

c *1961 with the combined mill on the right and the die department in front.*

b

c

intensive jobs. The variety of work done in small batches in the refinery and in the rolling mills meant that manual labour was still the most flexible and cheapest method, as far as Thomas Bolton and Sons Ltd were concerned.

The years before World War II saw the end of the Bolton family's involvement with the firm. Thomas Bolton died in 1937, and was replaced as chairman by G.H. Nisbett of BIC. Frank Bolton and George Rathbone resigned as directors in July 1937. In August Michael Bolton was asked to resign from the company, and was replaced as Staffordshire manager by Commander H.C. Anstey. This proved an unpopular move with the workforce, but the strong non-Bolton interests were determined to push this through. Edward Bolton continued as sales manager, and Hugh Toller became production manager, both with seats on the board. Then Edward Bolton died suddenly in early 1939, and the last link was broken between the management of the firm and the family.

As the war clouds loomed, two further developments were proposed. The first involved a further expansion of the domestic copper tube business, and the building of a new tube mill at Froghall at a possible cost of £250,000. This then became linked to an agreement between Thomas Bolton and Sons Ltd and British Insulated Cables to have joint use of the Loewy extrusion press which had been installed at Prescot. In fact the tube mill was not built until after the war. The agreement over joint use caused problems throughout the war, as the press was not working as well as the buyers had expected.

The second proposal, which was carried through, was to set up a comprehensive technical and research centre at Froghall, under the control of Dr W.E. Alkins, at a cost of over £10,000. This was built at the

A heavy drawbench finishing a bronze piston rod.

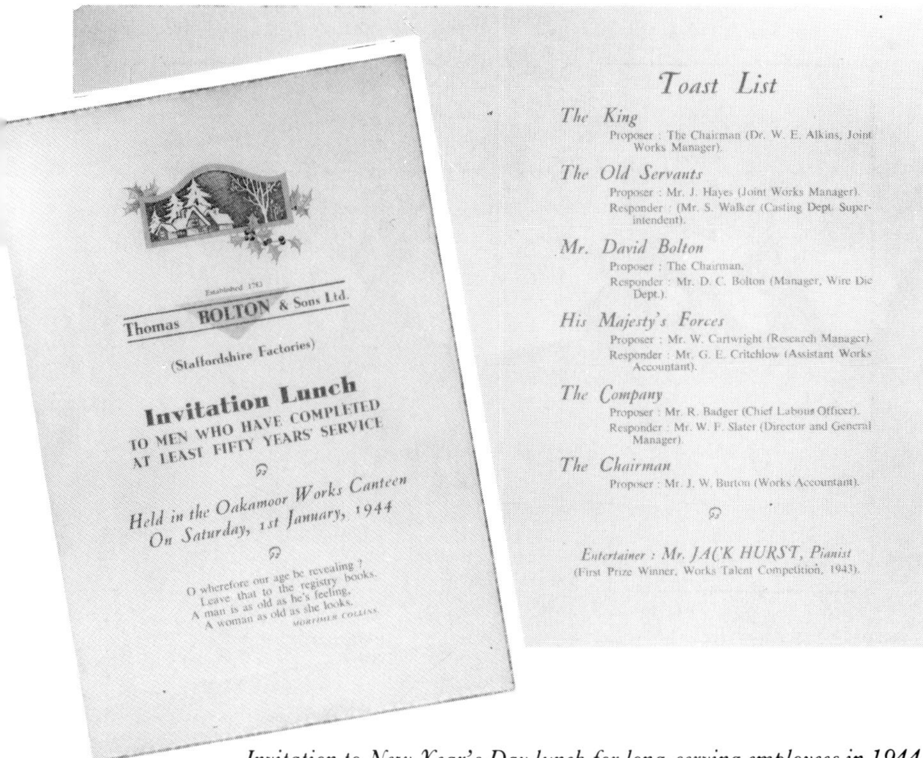

Established 1783

Thomas BOLTON & Sons Ltd.

(Staffordshire Factories)

Invitation Lunch

TO MEN WHO HAVE COMPLETED AT LEAST FIFTY YEARS' SERVICE

Held in the Oakamoor Works Canteen
On Saturday, 1st January, 1944

O wherefore our age be revealing ?
Leave that to the registry books.
A man is as old as he's feeling,
A woman as old as she looks.
MORTIMER COLLINS

Toast List

The King
Proposer : The Chairman (Dr. W. E. Alkins, Joint Works Manager).

The Old Servants
Proposer : Mr. J. Hayes (Joint Works Manager).
Responder : (Mr. S. Walker (Casting Dept. Superintendent).

Mr. David Bolton
Proposer : The Chairman.
Responder : Mr. D. C. Bolton (Manager, Wire Die Dept.).

His Majesty's Forces
Proposer : Mr. W. Cartwright (Research Manager).
Responder : Mr. G. E. Critchlow (Assistant Works Accountant).

The Company
Proposer : Mr. R. Badger (Chief Labour Officer).
Responder : Mr. W. F. Slater (Director and General Manager).

The Chairman
Proposer : Mr. J. W. Burton (Works Accountant).

Entertainer : Mr. JACK HURST, Pianist
(First Prize Winner, Works Talent Competition, 1943).

Invitation to New Year's Day lunch for long-serving employees in 1944.

beginning of the war in 1940, and provided invaluable service to the company.

With the outbreak of war, the company was once again turned over largely to munitions work. Parts for aeroplane engines, and the undercarriage assembly for the Lancaster bomber were among the main contributions made, along with the supply of vast quantities of copper shell-bands and cartridge cases. Naval requirements also led to extensions at Sutton and Widnes. At the same time, especially by 1943 when victory was seen to be only a matter of time, urgent consideration of the post-war situation was made at every meeting of the board. The firm, and British industry generally expected to be able to take advantage of the immediate post-war vacuum, in which they would face very little competition for their goods, and at the same time have a great demand for them during the period of reconstruction.

In 1946, the proposal to build a tube mill at Froghall was taken up again and expenditures of nearly £50,000 for plant and a further £35,500 for buildings were envisaged. This was proceeded with and the tube mill was in production by 1951.

By the early 1950s, Thomas Bolton and Sons Ltd's factories in Staffordshire were still the most important centre in the UK for the manufacture of copper and its alloys for electrical purposes. During the

Commutators and commutator segments for electric motors and generators.

A large machined generator end ring at Froghall.

Korean War price boom, the company's profits were doing well, rising to £614,830 in 1951, and a record £799,807 in 1952. A wide variety of products was being made by the 1,850 employees. Wire was still the most common form for high conductivity copper, along with strand, bar, rod, sheet, strip and foil, commutator segments and tubes. By now the vast majority of the high conductivity and oxygen-free high conductivity copper used was imported as wire-bars from Canada and Rhodesia. Some copper was still being reclaimed from scrap in the refinery at Froghall. From this copper numerous alloys were made, for which Thomas Bolton and Sons Ltd had now become well-known. Although other manufacturers made some of them, none produced the range of alloys in the variety of forms which the company could offer. Silver-bearing copper, under the trade-mark 'Combarloy', was made for items needing later heat-treatment. This had been developed under the late Thomas Bolton for applications in which a higher than normal softening temperature was needed. Cadmium copper and low-tin bronzes, mostly for trolley wire, tellurium copper for easier machining qualities, and chrome copper for spot-welding electrodes were some of the other alloys produced. These also were largely the fruits of the late chairman's interest in research and his active encouragement of it. In addition to these alloys, aluminium and phosphor bronzes were made, and a variety of brasses, including tin, silicon and aluminium brasses, as well as the basic copper/zinc brass in proportions ranging from 80/20 to 62/38.

Line and cable wire, that is wire of 0.128in diameter or less — the company made wire down to 0.004in — was made from ¼in rod rolled in the special mill which was the successor of the billet, or breaking-down mill of 1909. The larger gauges of wire, above 0.0128in, mostly for trolley and other contact wire, were made in the combined rolling mill installed before the war. This mill also produced bar and strip for commutator segments and other purposes. The Schloemann two-high reversing mill, which was the first stage of the treatment, had box-grooves for reduction and a flat for strip. The copper then passed through the intermediate and finishing trains. Over a twelve-month period, over 1,000 different sizes of rounds and flats would be made.

This did not exhaust the range of processes carried out on high conductivity copper by Thomas Bolton and Sons Ltd. The first extrusion press was installed in 1916, and this was replaced in the mid-1950s by a 1,800 ton Schloemann press, capable of taking billets up to 24in long. This produced a large amount of intermediates, notably brass rod, copper and brass strip and simple shapes such as tapered strip for commutator sections.

The company thus seemed poised for a successful post-war career. But changes, both external and internal, were to have far-reaching effects on the company's activities. Firstly came the end of the steam age on British Rail. Edward Bolton had warned in 1937 of the trend towards larger fire-boxes for main line locomotives, and the increased use of electric and diesel power for suburban lines and shunting. However, it was decided

SAMUEL PLATT LTD
SIZE 3

Sawing commutator strip to shape during World War II.

◀ *Drawing commutator strip on a block in the brass department at Froghall during World War II.*

◀ *Drawing commutator strip on a bench in the combined mill at Froghall during World War II.*

that nothing be done for the time being. By 1948 the company was finding that British Rail was asking for larger fire-boxes, as he had predicted, but the company was uncertain whether to invest in new plant without a clear idea of British Rail's intentions. In fact the Railway Executive, dominated by the steam lobby, continued to order a new standardised range of locomotives. As a result of this, a gradual and progressive changeover to diesel and electric traction did not take place, and steam, still king in 1955, was quickly phased out in the late 1950s and the early 1960s.

The sudden loss of much of the locomotive business, a major part of the company's non-electrical output since 1838, was compensated for to some extent by new work arising from electrification. Brush Electric ordered copper commutator segments and windings for the DC motors in the new electric locomotives, and Thomas Bolton and Sons Ltd also made bronze overhead 'hockey-stick' fittings for the Glasgow-Euston line. This work of course was done at Froghall, and there was no new alternative outlet for the Widnes rolling and plate mills. The sulphate works suffered similarly as the countries which formerly bought 'Mersey Brand' sulphate began to make their own.

The switch to jet aircraft by the Air Force and civil aviation in the 1950s meant that the Staffordshire factories lost their internal-combustion aero-engine business. The alloys involved in jet engines were outside Bolton's range. It was also decided to stop the manufacture of large diameter tubes, which were uneconomic.

Another break with the past was the decline of orders from the Post Office in the 1960s and 1970s. Bolton's bronze and later cadmium wire was employed on overhead telephone lines. But the Post Office was going over steadily to underground distribution using stranded soft copper wire, and so the long association between the firm and the Post Office declined.

After the formation of British Insulated Callender's Cables from British Insulated Cables and Callender's Cable and Construction Company in 1945, the Thomas Bolton group of companies had retained a considerable amount of independence. The main effects of the merger of Bolton's main shareholders were the introduction of a common pension scheme, and the changing of the financial year to fit in with the remainder of the combine. But as the changes outlined above began to have an effect, BICC began to rationalise their production lines. For Thomas Bolton and Sons Ltd the most historic consequence of this was the phasing-out of wire-making at the Staffordshire factories, putting an end to a tradition going back two and a quarter centuries. Wire-drawing was now to be concentrated in the former British Insulated Cables works at Prescot, and the Erith cable works set up by the Callenders.

In 1961, the remaining shares not in the hands of BICC were bought out, making Thomas Bolton and Sons Ltd a wholly-owned subsidiary. Following this, it was decided to finally carry out the closure of Oakamoor first considered before the war. The works had been gradually run down

Typical trade advertisements for Bolton's products for the electrical industry in the 1950s.

Oakamoor on Tuesday 10 September 1963. The big mill chimney says goodbye before being felled the following afternoon.

The end of an era at Oakamoor. On this historic site had been an iron forge as early as 1573, later an iron and tin-plate rolling mill, and copper and brass works since 1792. The site has been returned to nature and is now a pleasant recreation area.

in the 'fifties, and economically it was a necessary step for the company. The works were left for a while until under pressure from the Oakamoor villagers the derelict buildings were demolished in 1963. The site is now laid out as a recreation area, and little remains today to indicate the pioneering industrial activity which was carried on there.

Big changes were also overtaking the Lancashire factories. The remaining locomotive business was moved to Sutton. The main activity at Widnes was the copper sulphate works. But here also there was further rationalisation, involving the company's competitors. Originally there were three firms making copper sulphate: Thomas Bolton and Sons Ltd, Dennis, and McKechnie's. Dennis went out of business, and later an arrangement was made with McKechnie's whereby they made the copper shot, and this was turned into crystals by Boltons. Finally, the whole operation was taken over by McKechnie's, and Thomas Bolton and Sons Ltd ceased to have anything further to do with sulphate manufacture.

There were changes too at Sutton. By the mid-'fifties, the trade in copper printing rollers had died out, except for some high quality printing which was still done on copper sheets. The zinc plates, sold through the sole agency of SD Syndicate, were doing well. Unfortunately for Thomas Bolton and Sons Ltd's balance-sheet, however, most of the profits went to the middlemen. For this reason, BICC acquired SD Syndicate in 1963. Eventually, in 1974, Sutton was hived off from the Bolton companies, and became part of a new product group. The rolling mill, acquired by Alfred Bolton in 1894, continued with a diminishing output until its final closure in 1982. This copper mill was probably unique in the country, relying as it did on the manual skill of its workers to make copper plates of unusual shapes and sizes which could not be produced by more automatic methods. Among the last customers were whisky distillers who needed copper vessels with a complex curved shape most easily constructed from individually-made plates.

As a result of these changes, Froghall became the centre of the Thomas Bolton group inside BICC. Here, machinery very different to that still being used at Sutton was installed in 1961. This was the Loewy 1,700 ton extrusion press which was designed to give the maximum output of copper rod or strip of high quality, from a fully mechanised and automatic plant. In the same year, the cable strip mill was also built. This was to supply foil strip for a series of submarine telephone cables being made by Standard Telephone. The first trans-Atlantic telephone cable was not laid until 1956, a surprising 100 years after the first telegraph cable. The co-axial submarine cables being made by Standard Telephone were thus amongst the earliest of their kind. Only IMI and Thomas Bolton's had the facilities to make the coil but in the event IMI ran into difficulties and could not fulfil their side of the work. Thomas Bolton and Sons Ltd were thus carrying on appropriately where their predecessors had pioneered one hundred years before — at the bottom of the world's oceans.

Froghall from the air, early 1960s.

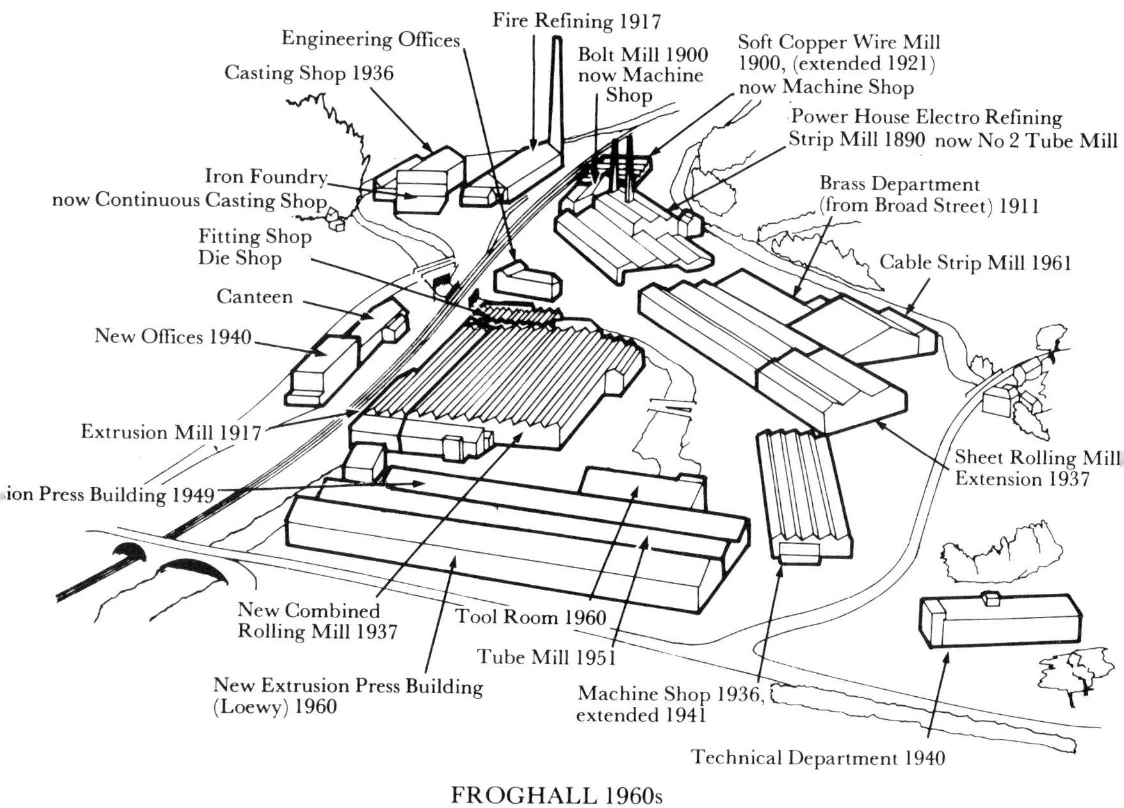

FROGHALL 1960s

Key to aerial photograph identifying some of the major buildings.

Thomas Bolton & Sons Ltd Today

Leaner than before, the Froghall works of Thomas Bolton & Sons Ltd concentrate on the 'Boltomet' range of coppers and copper alloys. Now the wire-drawing has mostly gone, high conductivity copper is made as strip, foil, bar, rod and busbars for power-stations and similar purposes, as commutator copper, and as a variety of machined parts. A wide range of copper alloys for electrical purposes are offered, including bronze, cadmium copper, chromium copper, the silver-bearing 'Combarloy', oxygen-free and sulphur copper. Aluminium and tin bronzes and cupro-nickels are made for mechanical purposes. Copper tubes in light gauges are made for the building industry. The manufacture of dies for drawing and extrusion of ferrous and non-ferrous metals from carbide products and new components and products for use in the oil, petro-chemical and steel industries complete the range.

Like all manufacturing companies, Thomas Bolton & Sons Ltd are facing very difficult times as the world recession deepens. In the slump before World War II, the company was able to develop a new range of products to compensate for lost business so now, in the present situation, new products are being developed to supersede the volume lost by the declone of consumption in general engineering industries.

Acknowledgements

The author wishes firstly to thank BICC and Thomas Bolton & Sons Ltd for the opportunity they have provided to write this book. Many people have given help, information and advice: the author wishes to mention Professor J.R Harris, Dr H. Holden and the BICC Archives Committee; Mr G. Tatton and Mr R. Shaw at Froghall; Mr R.M. Morgan at Prescot; Mr B. Mitchell at Wood Lane; Mrs J.A. Clarke in the Legal Dept, Bloomsbury Street; Mr T. Aston; Mr V.B. Hysel; Mr & Mrs Martin Bolton of Croxden Abbey; Mr N.O.B. Baker; Dr John Robey and Mr Lindsey Porter of Moorland Publishing; the late Mr H. Chester of Cheadle; Mr K. Davies of Liverpool Polytechnic; Mr T.V. Jackson of Pilkingtons; Birmingham Central Library Local Studies Department; Staffordshire County Record Office; Warrington Public Library; St Helens Public Library; British Non-Ferrous Metals Federation; British Aluminium Research Department; Kungliga Biblioteket, Stockholm; Mrs L. Shepherd and Professor G. Tucker. Finally, the author wishes to acknowledge the work of the late Dr Alkins in collecting material for a history of Thomas Bolton & Sons Ltd, which unfortunately he was unable to complete.

Many of the illustrations used in this book have come from Thomas Bolton & Sons Ltd at Froghall. Other illustrations have been provided by J. Plant, L. Porter, P. Wilson and Mr Martin Bolton, and grateful thanks are due for their use.

Bibliography

Before the Boltons

Manuscript Source

1 'Thomas Barker's Accounts', a rough notebook in the possession of Mrs Smith of Beelow Hill, near Cheadle.
2 Ledger accounts for New Spout 1761-6 and Old Spout 1769-72, in the possession of Thomas Bolton & Sons Ltd, Froghall.
3 Various deeds and indentures in the possession of Thomas Bolton & Sons Ltd at Froghall.
4 The Patten Papers in the Warrington Public Library.
5 John Ludwig Robsahm's *Dagbok en resa ig England* of 1761 in the Kungliga Biblioteket, Stockholm, gives a detailed description of the Cheadle brass works and Alton wire mill.
6 The Cheadle Brass Company's Minute Book, 1788-1831, in the possession of BICC, Prescot.
7 Wages Book for Oakamoor Mills 1844-65, Staffordshire County Record Office (SRO), D1406/1.

Printed Sources

Herbert Chester, *The Iron Valley*, 1979, and *Cheadle, Coal Town*, 1981.

Kenneth Davies, 'The Eighteenth Century Copper and Brass Industries of the Greenfield Valley', *Transactions of the Honourable Society of Cymmrodorion*, 1979, pp203-32.

Joan Day, *Bristol Brass*, 1973.

M.W. Greenslade and J.G. Jenkins, *A History of the County of Stafford*, 1967.

Henry Hamilton, *The English Brass and Copper Industries to 1800*, revised edition 1966.

J.R. Harris *The Copper Industry in Lancashire and North Wales, 1760-1815*, D Phil thesis, Manchester University, 1952.

John Percy, *Metallurgy-Copper; Zinc; Brass*, 1861.

Robert Plant, *History of Cheadle*, 1881.

John Robey, 'Copper Smelting in Derbyshire' *Bulletin of the Peak District Mines Historical Society*, vol 4, pt 5, June 1971.

John Robey and Lindsey Porter, *The Copper and Lead Mines of Ecton Hill*, 1972 (Moorland Publishing Co Ltd).

Samuel Timmins, ed, *Birmingham and the Midland Hardware District*, 1866.

The Boltons

Manuscript Sources

1 Alfred Sohier Bolton's Diaries, in nine notebooks, 1851-1901.
2 Francis Arthur Bolton's Diaries, in eleven notebooks, 1878-1902.
These are the most important single source of information for the nineteenth century. In the possession of Mr Martin Bolton.
3 Wire Strip and Rolling Account and Wire-Drawers Account, 1852-8, SRO.D1406/2.
4 Oakamoor Wire Production Book 1864-9, in the possession of Mr P. Wilson.
5 Will of Samuel Bolton of Walsall, Lichfield Diocesan Archive.
6 Wills of Richard, Elizabeth, Thomas and Alfred Sohier Bolton Public Record Office (PRO), Prob 11.
7 Birmingham Rate Books for Edmund St, Paradise St, Newhall St, New St, and Broad Street, 1780-1850, Local Studies Dept, Birmingham Central Library.
8 Various Deeds and Indentures at Froghall (F) and at BICC Head Office in Bloomsbury St, London (B):
Partnership Deed of Thomas Bolton & Sons, 30 June 1853 (F).
Indenture of sale of Thomas Bolton & Sons to Thomas Bolton & Sons Ltd, 10 June 1902 (F).
Agreement for sale and purchase of Oakamoor, 1 March 1852 (B).
Agreement for the purchase of the Oakamoor freehold, 1 May 1852 (B).
Conveyance of Froghall site, 25 August 1890 (B).
Partnership Deed for Francis Arthur Bolton, 2 May 1892 (B).
9 Letter of Thomas Bolton, 16 March 1934, re Widnes 1881-1900 (F).
10 The Sohier Papers, Massachusetts Historical Society, Boston, Mass.

Printed Sources

Arris's Gazette (Birmingham Newspaper) 1782-95.
T.C. Barker & J.R. Harris, *A Merseyside Town in the Industrial Revolution: St Helens 1750-1900*, 1959.
C.H. Beale, *Memorials of the Old Meeting House and Burial Ground*, 1882.
Birmingham and District Trade Directories, 1760-1855.
B.C. Blake-Coleman, *The Rise of Copper Wire, its manufacture and use*, M Phil thesis, Open University, 1980.
A.S. Bolton, Report of his contribution to discussion after W.H. Preece's paper to the Institution of Civil Engineers in 1883 (see below under Preece). This is the only published source from the Bolton family in the nineteenth century.
H.C. Bolton & R.P. Bolton, *The Family of Bolton*, New York 1895.
(British Aluminium), *History of the British Aluminium Company 1894-1955*, 1955.
R. Chadwick, 'The Working of Metals' in the C. Singer, et al, *History of Technology*, vol 5, 1958.
R.K. Dent, *The Making of Birmingham*, 1894.
H. Dunell, *British Wire-Drawing and Wire-Working Machinery*, 1925.
P. Dunsheath, *History of Electrical Engineering*, 1962.
M.B. Hall, 'The Strange Case of Aluminium', *History of Technology*, 1976.
D.W.F. Hardie, *A History of the Chemical Industry in Widnes*, 1950.

J.G. Jenkins, contribution on Engineering in Greenslade and Jenkins, *Victoria History of the County of Stafford*, 1967, vol 2.

Langford's, *A Century of Birmingham Life 1741-1841*, 1839.

Lardner's, *Cabinet Encyclopaedia*, III, p185, 1834.

R.M. Morgan, *Callender's, 1882-1945*, BICC, 1982.

R. Plant, *History of Cheadle*, 1881.

Report of the Joint Committee appointed by the Lords of the Committee of Privy Council for Trade and the Atlantic Telegraph Company. (A House of Commons paper published in 1861.)

J.B. Smith, *A Treatise on Wire*, 1891.

S. Timmins, ed, *Birmingham and the Midland Hardware District*, 1866.

R.R. Tomey, *Vivian and Sons 1809-1924*, D Phil thesis, University of Wales, 1979.

Walford's, *County Families*, 1896.

Articles

Edward Bolton, 'Manufacture of Brass Wire', a paper for the *Institute of Mechanical Engineers Graduate Association*, 1906.

(Deputy Editor) 'The Copper Works of Messrs Bolton & Sons', *Electrical Review*, vol 40 no 1008, 19 March 1897, pp385-91.

J.C. Duncan, 'Victorian Telecommunications', *Journal of Post Office Electrical Engineers*, vol 74, April 1981, pp52-4.

W.H. Preece 'On Electrical Conductors', in the *Minutes of Proceedings of the Institution of Civil Engineers*, vol lxxv, 4 December 1883, pp63-78.

W.H. Preece *Journal of the Society of Arts*, vol xxxvi, 1888.

Willoughby-Smith, 'A Resumé of the Earlier Days of Electric Telegraphy', *Journal of the Society of Telegraph Engineers and Electricians* vol x, no 38, 22 September 1881.

Articles on the Atlantic Telegraph in *The Engineer*, 21 November and 19 December 1856; 30 January and 12 June 1857; 28 September 1866.

Anon, 'Messrs Bolton's Oakamoor Wire Works' *Uttoxeter Advertiser*, 2 June 1909.

Anon, 'The Staffordshire Works of Thomas Bolton & Sons Ltd', *The Ironmonger*, 21 July 1934.

Anon, *Wire Industry*, Special Overseas Edition, July 1937.

Index